JANICE PRATT VANCLEAVE

101 Activities to Make
Science Education Easy and Enjoyable

Teaching
the Fun
of Physics

PRENTICE
H A L L
PRESS

New York   London   Toronto   Sydney   Tokyo   Singapore

Prentice Hall Press
15 Columbus Circle
New York, New York 10023

Copyright © 1985 by Prentice Hall Press
A Division of Simon & Schuster, Inc.

Published in 1987 by Prentice Hall Press
Originally published by Prentice-Hall, Inc.
Cover and interior illustrations by April Blair Stewart

PRENTICE HALL PRESS and colophon are registered
trademarks of Simon & Schuster Inc.

Library of Congress Cataloging-in-Publication Data

VanCleave, Janice Pratt.
  Teaching the fun of physics.

  (The Prentice-Hall science education series)

  Bibliography: p.
  Includes index.
  1. Physics—Study and teaching (Elementary)
2. Physics—Experiments.   I. Title   II. Series.
QC33.V34   1985   372.3'5   85-3491
ISBN 0-13-892415-5

Manufactured in the United States of America

10   9   8   7

In memory of my mother,
Wilma Frankie Pratt (1923–1978),
and nephew Christopher E. Pratt (1964–1980)

Contents

Foreword

Fun! This book is fun to read and fun to use. Youngsters and adults both will delight in boiling water with ice or playing with a coffee can that returns after being rolled away. Here is magic that comes not from tricks and illusions but from the majesty of nature.

Yes, *Teaching the Fun of Physics* is educational, too. Physics is the foundation of modern science, our ultimate tool in understanding the laws that govern the universe. And just as we see the ocean in a drop of water, so we see the laws of optics in a handmade periscope and in a picture projected on a wall by a magnifying lens.

The secret to more rapid learning and better memory is simple: Change your view of information to a form that is easily absorbed and retained by your mind. Janice VanCleave did this for us in *Teaching the Fun of Physics*. She describes the results of each activity and gives the scientific explanation of these results in an easy, clear style. Each concept is backed by a concrete experience so that learning becomes as natural and pleasurable as play.

Despite their high impact, the experiments are simple and easy to perform. Most of the equipment can be found in the household. Jars, string, paper, and tape are among the staples. When occasional magnets or lenses are needed, they are inexpensive and easy to obtain. Personally, I have an incredible lack of talent for performing experiments. (As a theoretical physicist, I do my work with pen and paper.) The happy fact is that I am now having a grand time adding spice to my university lectures with experiments from this book. Even I can do them—and they work!

Robert W. Finkel
Chairman, Department of Physics
St. John's University, New York

Preface

This is an elementary science experiment book designed to teach the fun of physics. The experiments were selected on their merit to demonstrate that science *is* fun, exciting, and at times a bit mystical. The experiments—all 101 of them—are unique in their ability to present scientific information in a way that makes learning an enjoyable experience.

To ensure success in performing each experiment, detailed step-by-step instructions are given along with illustrations. All the activities have been pretested, and they work if the directions are specifically followed—exactly in the steps described.

To further guide the experimenter, expected results are given. Corrective procedures are described for the few experiments in which predictable errors might occur.

Another special feature of this book is the scientific explanation of why the results were achieved. The reader will be pleasantly surprised to find this section written in understandable terms.

Some equipment and supplies are needed, but in all the experiments the necessary items can be easily obtained. All of the materials can be found around the home or are readily accessible at a local grocery or variety store.

All the experiments in this book could be used in designing a science-fair project. A special section is included (see appendix) pointing out how a topic can be developed into a suitable project. A few specific topics have been chosen and ideas given on the needed research, construction of models, and experimentations needed to prepare for entering the project in a science fair. The appendix also contains display ideas. Methods of designing different types of projects and invaluable display ideas are described.

This book of 101 experiments contains experiments related to four different science curriculums—biology, physics, geology, and chemistry. Each experiment has step-by-step instructions and illustrations, expected results, and a scientific explanation in understandable terms.

The book was written to provide workable experiments. Because of the use of easily accessible supplies, all the experiments can be performed with rewarding success for the experimenter.

The science fair section gives more than just a list of topics. It shows by example how to take an idea and develop it into a possible award-winning project.

The objective of this book is that by making the learning of scientific information enjoyable, there will be a desire to learn more about science.

Acknowledgments I wish to express my appreciation to Crystal Settle, a director of the continuing education department of Westark Community College, Ft. Smith, Arkansas. As a result of the elementary level enrichment program called "The Magic of Science" that Ms. Settle asked me to design and teach, my material was noticed and subsequently published.

A special note of gratitude to my friend and fellow author Joyce James for the much-needed information and support.

This book has been in the making for years, and hundreds of elementary, high school, and college students have performed the pre-tested experiments. I want to thank the members of my family who willingly and at times unwillingly volunteered their time to do the pre-testing.

My sister and brother-in-law, Dianne and Kenneth Fleming, did not participate in the testings but with much zeal allowed me to borrow their children, Kymie, Kenneth, and Robert, who did.

My brother and his wife, Dennis and Brenda Pratt, also have been most generous in allowing me to work with their children, Chris, Carol, and Bobby Raymond.

My nieces and nephews were usually privileged to enjoy the fun of science only on weekends and holidays. However, my own children, Ginger, Russell, and David, could experience the joy more often and especially if they brought home a new friend who had not seen the latest demonstration. I do appreciate their patience.

Russell's marriage to Ginger and the birth of my granddaughter Kimberly, alias BoBo, was especially thoughtful of him since it has given me two fresh new experimenters to work with.

My mom and dad, Frankie and Raymond Pratt, certainly deserve credit for any scientific knowledge that I acquired while in school, as they thwarted my plan of dropping out at the age of fourteen.

Patsy and Marvin VanCleave have always been encouraging, but their prime contribution was Wade, their son and my husband. Wade's scientific interest is restricted to the procedure in catching a trophy bass, but because of the love he has for me he read every experiment and offered suggestions on how to make them more understandable. In addition to his evaluating, editing, and typing of the manuscript, his kindness kept me sane when I typed half the book incorrectly and drew 200 unacceptable illustrations before reading the book *How to Write a Book*. I am truly grateful for his support.

Introduction

The science of physics is a looking, feeling, seeing, hearing, and tasting experience. It is a way of asking a question, then formulating an answer from information derived from experimentation and research. As a result of such scientific investigations, physicists—and your students—gain insight into how things work, develop, and interact.

A scientist is a person who uses the scientific approach for problem solving. This does not require a Ph.D. in physics or a genius-level I.Q., only an inquisitive mind, a desire to learn, and an understanding of the scientific approach, which is:

1. *Identifying a problem:* Before an experiment can be designed or research data collected from printed literature, the problem has to be identified. Stating the problem in question form aids in its clarification and sets limits on the needed research.

2. *Formulating a hypothesis:* A plan of action must be drawn up in order to solve the problem. The experimenter will collect known facts relating to the question from published materials in order to make an educated guess as to the answer to the question. This educated guess, known as a hypothesis, may be right or wrong. A wrong hypothesis is important in that it eliminates a possible answer; with successive trials and errors the correct answer will be found.

3. *Experimenting:* An experiment is chosen or designed to test the hypothesis.

4. *Collecting data:* (a) All results obtained from experimentation are recorded in an orderly form. At this point, every observation should be recorded. (b) Information collected from current publications can be added to experimental data. This information can enable the researcher to solve the problem without doing the experiment if previous work has been done in the area.

5. *Writing the conclusion:* This is a statement that summarizes the experiment. It should describe what was expected to happen and what actually happened. If the expected and actual outcomes are not the same, an explanation as to why they are different is given. It is from the conclusion that the answer to the problem can be found or another hypothesis formulated for future testings.

Learning the laws of physics, and productive science education in general, requires firsthand experiences. Learning to investigate is an integral part of education which leads to the development of problem-solving skills. These are applicable in learning to identify, evaluate, and solve problems outside the realm of science. Training to observe helps one in all aspects of life, from mastering proper driving skills to writing a term paper, where the skills of collecting data, classifying it, and evaluating related materials is imperative.

The idea that physics—or any other science—must be restricted to a laboratory containing sophisticated and mysterious equipment being manipulated by a person smocked in a white coat needs to be replaced with the thought that science can truly be for everyone. Many young potential scientists view physics as a difficult, complicated, and mysterious field totally inaccessible to them. Modern technology is too much a part of the world around us for one to shy away from its development. This is an era of electronic computers, space exploration, and nuclear research. Advancements from technological spinoffs make life more comfortable, more entertaining, and longer. This offers a rewarding and promising future for all and an increasing number of exciting careers in science.

There is always a starting point. Einstein did not decide to develop the theory of relativity on the spur of the moment. Much time and energy were exerted by Benjamin Franklin as he designed and carried out experiments relating to the properties of electricity. To obtain successful results and rewarding experiences in any endeavour, one needs to understand the basic concepts and procedures. This is as true in physics as it is in any field of study or problem-solving situation. Frustration and a feeling of inadequacy result from skipping the foundational materials and proceeding to the advanced, more complicated theories when studying science.

This book is designed to be a starting point for a person interested in physics or any area of science. It is complete in itself for one seeking a basic study in physics and a beginning step for studies in more advanced science curriculums. The material was selected to whet the appetite of students for further research into the world of science and specifically the area of physics. The reader will be pleased to find that much of the scientific terminology has been omitted and replaced with explanations in everyday language.

Atomic, kinetic, quantum, and other technical theories found in most formal studies of physics have been left out. Topics dealing with things that can be touched and watched, and which provide entertaining experiences, have been included. The four basic fields of science—biology, chemistry, geology, and physics—are subdivided into innumerable branches. Physics experiments alone are represented in this book, and even then only a touch of available physics information is included.

Physics is a study of the physical universe, dealing with motion, forces, work, energy, heat, sound, magnetism, electricity, and other relationships between matter and energy. It is a science that is applicable to everyday experiences.

With the present scientific advancements in physics there is an inclination to think that everything has been discovered. Lord Kelvin, a renowned physicist, said at the turn of the century that everything relating to physics had been discovered and that only a few minor points needed to be worked out. These minor points turned out to be the development of theories about x-rays, quantum mechanics, relativity, and many more topics. Physics is just now unfolding, and there is always the possibility that with the technological improvements new laws and theories will replace those currently accepted as being true. The building of cyclotrons, nuclear reactors, and other instruments aided physicists to better study the behavior of atomic particles, and many subatomic particles have been found. Prior to these scientific advancements, it was believed that the atom contained only protons, electrons, and neutrons. Students of science need to approach their studies with the idea that much still needs to be explained and discovered, and possibly they will be the ones to uncover this new material.

This is basically an elementary science experiment book, designed for use at the lower levels due to the simplicity of the wording but applicable to higher levels in its basic demonstration of more complex theories. It includes help for teachers, particularly those in elementary school, and college students preparing to teach. Children as well as teachers need no previous background in science, no costly equipment; only a willingness to try new things, seek answers, and generally experience instead of just accepting ideas from others.

Teaching the *fun* of science is the main theme of the book. The experiments were selected on the merit of their ability to demonstrate that physics is fun, exciting, and at times a bit mystical. The experiments are unique in their ability to present scientific information in such a way as to make learning an enjoyable experience. Upon completion of all the activities the experimenter will have developed a familiarity with the topics of buoyancy, gas pressure, Bernoulli's principle, light, heat, electricity, gravity, inertia, surface tension, and wave motion.

Each experiment is designed for individual work. Parents and/ or teachers need to allow the child to perform the experiments with minimum help or instructions. The learning process is slow, and actual hands-ons, do-it-yourself activities increase the learning skills. Young experimenters often appear to an adult observer to be "messing around" or "messing up," but through these experiences the child can learn to make observations, to make decisions, and about the need to follow an experiment through to its conclusion. Each experiment tells the child what the expected results are, so a "foul-up" is quickly recognized and the need to start over or make corrections is immediately apparent.

To insure success in performing each experiment, detailed, step-by-step instructions are given, along with illustrations. All activities have been pretested, and they work as long as the directions are specifically followed. Substitution for the equipment called for can alter the described results.

To further guide the experimenter, expected results are given along with corrective procedures for the few experiments in which predictable errors might occur. To be aware of the expected outcome of an experiment can, as has been previously suggested, allow the experimenters to recognize errors in their work, but it is also a reinforcement for correct procedures.

Another special feature of the book is the scientific explanations of the results and why they were achieved. The reader will be pleasantly surprised to find this section written in understandable terms. When scientific expressions are necessary, they are explained when first introduced, but repeated definitions are not given. A glossary of all scientific terms, explained in simple terminology, is included.

Equipment and supplies are needed, but in all of the experiments the necessary items are of a nonscientific nature. All of the materials can be found around the home or are readily obtainable at a local grocery or variety store. The use of common materials emphasizes that science is and can be a part of our daily lives.

All of the experiments can be used to design a science fair project. A special section is included which discusses how a topic can be developed into a suitable project. A few specific topics have been chosen, and ideas are given on the research, construction of models, and experimentation needed to prepare for entering these projects in a science fair. Methods for designing different types of projects and invaluable display ideas are described.

There is no particular order for performing the experiments, but it is advisable to do each section in sequence, as there is some buildup of information, and scientific terms introduced are defined in the first experiment.

Success in any endeavor is a direct result of preparing, following instructions, and performing properly. The reader will have best results if he or she adheres to the following instructions and meticulously follows the instructional steps given for each experiment.

General Instructions for the Reader

1. *Read First:* Read each experiment before starting. This allows you to recognize what the exercise is about and to identify the problem or purpose.

2. *Collect Materials:* Tasks are made easier by having all needed items ready for instant use. This trains you in preparing not only for a science experiment but for any job that might be started. A scientist must analyze a problem and decide how to solve it, but before experimentation can begin the needed supplies must be collected and organized.

3. *Experiment:* This is what you want to do immediately, but the results are always unrewarding if you are not sure of what is to be done. Follow each step slowly and carefully and do not skip steps or add your own. Safety is of utmost importance, and by reading any experiment completely before starting, and by following the instructions exactly, you can feel confident that no unexpected results will occur.

4. *Observe:* Your results should be the same as those given for each experiment. If they are not, carefully reread the instructions and start over from the first step.

5. *Collect Data:* Keep a bound notebook of results, and make note of difficulties you experience along with a conclusion for each experiment. Organize material into topics, and add to it as other related experiments are completed.

The prime objective of this book is to make the learning of scientific information enjoyable so that there will be a desire to learn more about science. This objective can be met because each experiment has been tested and retested to insure that the experimenter will be rewarded with the expected results.

Physics can be one of the more exciting and "fun" studies of science. Many magical demonstrations can be explained by the laws of physics. Most toys are designed with physics applications. After completing this book students may enjoy the things around them more because of the increase in their observational skills and a new knowledge of the whys and hows of the scientific world.

1
Buoyancy

1. Floating Egg

- *Pour ½ cup of water into a clear glass.*
- *Carefully lower a small raw egg into the water.*

Results The egg sinks to the bottom.

- *Carefully stir table salt into the water, being careful not to break the egg. It may take at least 2 tablespoons of salt, depending on the weight of your egg, before the egg floats on the surface of the liquid.*

Results The water looks cloudy because of the excess salt. This cloudiness will clear up in a few minutes, but you will be able to see the egg floating on the surface as soon as the salt is stirred in the water.

- *Add ½ cup of fresh water to the glass containing the salt water. This must be done very slowly and carefully to avoid mixing the fresh and salty water. Do this by holding a large spoon just above the surface of the salt water and against the side of the glass. Pour the water into the spoon and let it slowly pour out into the glass.*
- *Remove the spoon.*

9

Results The egg floats in the center of the liquid.

Why? All objects that float do so as a result of the upward force exerted by the liquid. This force is called the *buoyant force.* The degree of this force depends on the weight of the liquid and the size of the submerged object. Archimedes, a Greek scientist, discovered that the buoyant force is equal to the weight of the liquid displaced by the object. (*Displaced* means that the liquid is pushed out of the way.)

The buoyant force on the egg in the fresh water is equal to the weight of the water displaced by the egg. This force is less than the weight of the egg, so it sinks. The egg in salty water floats because the weight of the salt water displaced is greater than the weight of the egg. The egg floats between the salt and fresh water because the egg falls through the fresh water layer and is supported on top of the second, heavier layer of salt water.

10

2. Jumping Rocks

■ *Fill an 8-ounce, clear drinking glass three quarters full of water.*
■ *Add 1 teaspoon of baking soda to the water and stir.*
■ *Make 10 tiny balls of clay. The size of the clay balls should be about twice the size of an air rifle shot commonly called a "B-B."*
■ *Measure 5 tablespoons of vinegar into a cup.*
■ *Pour the vinegar into the water. Do not stir.*
■ *Immediately drop the clay balls into the glass of water one at a time.*
■ *Wait!*

Results The clay balls will start collecting bubbles, and some will start rising to the surface. As soon as the balls hit the surface, they spin over and fall to the bottom of the glass. Some will fall part of the way and start to rise again.

Why? When the vinegar is added to the baking soda, carbon dioxide gas is produced. The bubbles of carbon dioxide collect on the balls of clay. The bubbles act like tiny inner tubes and make the clay balls light enough to float. When the balls hit the surface of the water, the bubbles are knocked loose and the clay is too heavy to float without them. The balls that sink part of the way and then resurface have landed on bubbles that push them up. The carbon dioxide gas attached to the clay pieces makes them buoyant.

3. The Diver

■ *Fill a small-mouthed gallon jug to overflowing with water.*
■ *Partially fill a glass eyedropper with water.*
■ *Put the eyedropper into the water-filled jug. It should float.*
■ *Place the palm of your hand over the mouth of the jug and push down.*

Results The eyedropper sinks when you push down on the water and rises when you release the pressure on the water.

Problems That Might Occur
■ The eyedropper does not sink. To correct this problem, add more water to the eyedropper.
■ The eyedropper sank but never surfaced. To correct this problem, remove the eyedropper and squeeze some of the water out of it.

You can regulate the ease with which the dropper sinks and surfaces by the amount of water in the eyedropper.

Why? When you push down on the water, more water is forced into the eyedropper. This causes it to be heavier, and it sinks. Releasing the pressure on the surface of the water allows the water that was forced into the eyedropper to come out. The dropper then rises. With less water, the dropper is more buoyant.

4. Weigh Your Hand

- *Fill a bucket three quarters full of water.*
- *Place the bucket on a food scale, and record the weight.*
- *Put your hand in the water, and record the new reading on the scale.*
- *The difference in the two scale readings will be the weight of your hand.*

Why? The weight of the human body is close to the weight of an equal amount of water. It is not exactly the same, but it's close enough to compare.

You saw the water rise in the bucket when you put your hand in it. The size of your hand is equal to the amount of water that was pushed up. Since the weight of water is similar to the weight of your body, putting your hand in the water is like adding more water.

According to Archimede's principle, the buoyant force is equal to the weight of the water displaced. Your hand pushes water out of the way, and the increased scale reading indicates the weight of the water displaced. This is also a measure of the buoyant force on your hand.

5. Treasure Divers

■ *Fill an 8-ounce, clear drinking glass three-quarters full of water.*
■ *Add 1 teaspoon of baking soda to the water, and stir.*
■ *Pour 2 tablespoons of vinegar into the water <u>without stirring</u>.*
■ *Immediately add ¼ teaspoon of coffee grounds to the water. (The grains left in the filter after coffee has been made are called coffee grounds.)*

Results Each coffee grain dives and surfaces with what appears to be a crystal ball.

Why? The crystal ball "treasure" is actually a bubble of carbon dioxide gas that has attached itself to the coffee grain. The bubble acts like a tiny balloon, making the grain buoyant enough to rise to the surface. When the grain with the attached bubble hits the water's surface, the bubble is knocked loose, and the grain falls back to the bottom, or lands on another bubble partway down and again surfaces. These bubbles of carbon dioxide are produced by the chemical reaction of the vinegar and baking soda.

2
Gas Pressure

6. Sonic Egg

■ *Peel a hard-boiled egg.*
■ *Cut 2 strips of newspaper (about 2 inches by 10 inches each).*
■ *Hold the 2 paper strips together, and twist them. You are making a safe paper torch. If you do not twist the paper, it will burn too quickly and could burn your fingers.*
■ *Light one end of the twisted paper with a match.*
■ *Drop the burning paper into a gallon jar. The mouth of the jar must be just slightly smaller than the egg so that the egg will not fall in when sitting in the mouth.*
■ *When the paper stops burning, <u>immediately</u> place the egg in the mouth of the jar.*
■ *Wait and watch!*

Results You will see the egg move slowly at first into the jar, and then—BOOM!—it will seem to explode into the jar.

Why? As the paper burns, it removes most of the oxygen from the air inside the jar. The lack of this gas reduces the pressure inside. The pressure of the air outside the jar is so much greater that the egg is shoved into the jar.

Try This Instead of using a hard-boiled egg, cut a cardboard circle the size of the opening. Wet the cardboard circle, and hold it while you again thrust a burning piece of paper into the jar. As soon as the paper stops burning, place the wet cardboard over the mouth of the jar. In a few seconds the circle will be pushed into the jar just as the egg was.

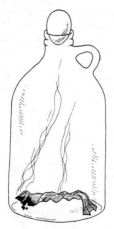

7. Barometer

- *Split a round 12-inch balloon in half.*
- *Stretch half of the balloon across the mouth of a quart jar, and secure it with a rubber band.*
- *Cut a drinking straw so that one end is pointed.*
- *Glue the uncut end of the straw horizontally to the center of the stretched balloon.*
- *Secure a ruler to another quart jar with a rubber band. The ruler must stand upright with the 1-centimeter measurement at the bottom. (It's best to use the metric scale to measure the changes in pressure, since the marks are closer together and you can measure small changes.)*
- *Position the 2 jars so that the end of the straw points to the metric markings on the ruler. Do not allow the straw to touch the ruler.*
- *Record the measurement that the straw points to each day at the same time for one week.*

Results The pointer will be in different positions on some of the days.

Why? As the air pressure outside the jar increases, it pushes down on the balloon, and the straw rises. A decrease in outside pressure causes the air inside the jar to push up, and the straw pointer is lowered.

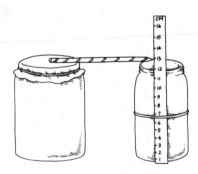

8. Two Worse Than One?

- *Place 1 straw into a glass of water.*
- *Suck on the straw with your mouth.*

Results Water moves from the glass up the straw and into your mouth.

Why? Sucking on the straw removes most of the air from your mouth and the inside of the straw. The air pressure outside the straw pushing down on the surface of the water is greater than the pressure inside the straw and your mouth. The difference in pressure causes the water to be pushed up the straw and into your mouth.

Try This
- *Place 2 straws into your mouth.*
- *Leave the end of one in the air, and place the other in the glass of water.*
- *Suck on both straws at the same time.*

Results You will find it difficult, if not impossible, to drink the water.

Why? One straw continues to bring in air, which keeps the pressure inside your mouth about the same as that outside your mouth.

9. Squirting Bottle

■ Fill a glass soda bottle three quarters full with water.
■ Use modeling clay to seal a drinking straw in the mouth of the bottle, with the end of the straw below the water's surface.
■ Hold the clay with your fingers as you blow through the straw into the water. (Note: The clay has not formed a seal if you can continue to blow air bubbles into the water for some time. You should be able to blow a few bubbles, then no more.)
■ Quickly move back from the bottle when you can no longer force air bubbles into the water.

Results The water inside the bottle squirts out the top of the straw.

Why? The air inside the closed bottle becomes compressed when you force excess air into it. Removing your mouth from the straw allows the compressed air to expand and move up the straw, and with it comes water.

Try This Use a bigger bottle so that there is more air to be compressed.

Results The water will squirt much higher.

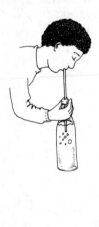

20

10. Inverted Glass

■ *Fill a drinking glass with water to overflowing.*
■ *Place an index card or piece of stiff cardboard over the mouth of the glass. The paper must cover the entire opening of the glass.*
■ *Hold the card in place as you turn the glass upside down. (You might want to do this over a sink or large pan, just in case the water spills.)*
■ *Remove your hand from the paper.*

Results The card stays over the mouth of the glass, and the water stays inside.

Why? The air pressure pushing up on the card is greater than the pressure of the water pushing down. Air pushes with a force of about 14.7 pounds per square inch. This is enough to hold up the water.

Try This Completely open one end of a metal can. Make a small hole with a nail on the side near the closed end. Follow the instructions for the inverted glass, but when you start to turn the can over cover the hole with your finger. Challenge a friend to do the experiment.

Results You will be able to do the experiment as long as the hole is covered, but your friend will not if the hole is open.

Why? When the hole is open, air will go into the can and push the water out. Air entering the can increases the pressure inside the can.

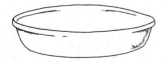

11. A Bottle Thermometer

- Wrap modeling clay around the middle of a drinking straw.
- Fill a cup one quarter full of water, and add enough blue food coloring to make the water a dark blue.
- Dip one end of the straw into the blue water.
- While the straw is in the water, place your index finger over the open end of the straw.
- Continue to hold your finger over the end of the straw, and insert the free end into a soda bottle. Before removing your finger from the end, push the clay around the opening of the bottle to seal it.
- Remove your finger from the end. The water bubble will move up the straw but should not come out the top. If it does, try again.
- Your thermometer will look like the one in the diagram.
- Place your hands around the soda bottle.

Results The colored bubble in the straw will start to rise to the top.

Why? The heat from your hands increases the temperature inside the bottle. The air in the bottle expands when heated, and the expanding gas pushes up on the colored water bubble. Most thermometers have a liquid such as mercury in them, and as the liquid is heated it expands and rises in the tube. This experiment shows you how a gas expansion thermometer works. Remember that the water bubble is not expanding but is being pushed up by the expanding air.

Try This Before the water bubble comes out of the top of the straw, run cold water over the outside of the bottle.

Results The colored bubble will move down the straw.

Why? The cold water causes the air inside the bottle to cool and contract. The air outside has more pressure and pushes the bubble down.

12. Popping Coin

■ *Place an empty, dry soda bottle in the freezer for at least 15 minutes.*
■ *Remove the bottle from the freezer, and immediately cover the mouth with a wet dime. The dime and the bottle must not have any chips on their edges.*
■ *Wet your fingers, and allow the excess water to drip around the edge of the dime. Be sure to get a layer of water around the edge, since the water acts as a seal. Wait!*

Results The dime will start to make a popping sound as it rises up on one side and falls back into place. This continues.

Why? Placing the bottle in the freezer allows it to fill with more air than it holds at room temperature. Cold air contracts and takes up less space. Once outside the freezer, the excess cold air inside the bottle starts to heat up and expand. This expanding gas pushes up with enough pressure to lift the dime on one edge. The dime falls back down as the gas escapes.

Try This Place your hands around the bottle to speed up the heating, and the dime will pop much faster.

13. Expanding Balloon

- *Place an empty, dry glass soda bottle in the freezer for at least 15 minutes.*
- *Remove the bottle from the freezer, and <u>immediately</u> cover the mouth with a balloon.*
- *Wait!*

Results The balloon will start to inflate. The size it reaches will depend on the strength of the rubber that the balloon is made of.

Why? Placing the bottle in the freezer allows it to fill with more air than it would at room temperature. Cold air contracts and takes up less space. The excess cold air inside the bottle starts to heat up and expand when the bottle is sitting in the room. The expanding air moves into the balloon and inflates it until the gas stops expanding. The pressure exerted by the balloon's material also stops it from expanding if it is not very elastic.

Try This Once the balloon has expanded to its maximum size at room temperature, put the bottle with the balloon still inflated inside the freezer. Check on it periodically, and observe the change in the balloon size.

Results The balloon deflates.

Why? The air cools and contracts inside the balloon and the bottle.

14. Fountain Machine

■ *Place one end of a straw into the mouth of a clean, dry, glass soda bottle.*
■ *Secure the straw, and seal the opening with modeling clay.*
■ *Fill a quart jar about three quarters full with water.*
■ *Add drops of food coloring and stir until a dark-colored water solution is formed.*
■ *Turn the bottle over so that the free end of the straw is in the jar of colored water. Keep the end of the straw at least 1 inch below the water's surface.*
■ *Place a hot, wet rag on the top of the inverted bottle.*

Results Bubbles will form in the colored water.

Why? The air in the bottle is heated by the rag and expands. As the air expands, it moves out of the straw, forming gas bubbles in the water.

■ *Remove the hot rag, and replace it with a wet, cold one. The cold rag can be prepared by placing the wet rag in the freezer for 5 minutes prior to this part of the procedure.*

Results Colored water from the jar will move up the straw and into the soda bottle.

Why? The air inside the bottle contracts when cooled. The cold rag causes this cooling and contraction of the air. The result of the contraction is that the air pressure inside the bottle is less than that pushing down on the surface of the colored water. The greater pressure on the water forces it up the straw.

15. Collapsing Jug

■ *Fill a thin, plastic gallon jug half full of hot water from the faucet.*
■ *Allow the water to stay in the open jug for 1 minute.*
■ *Pour the water out, and immediately screw on the cap.*
■ *Wait!*

Results The jug collapses.

Why? The hot water heats the air above it. The heated air expands, and some moves out of the open jug. When the water is removed and the jug closed, the air that is left starts to cool and contract. Since there is less air inside the jug, it has less pressure than the air on the outside. It is the outside pressure that pushes the jug in.

Try This For a more dramatic effect use the following procedure:

■ *Pour 1 cup of water into a clean gallon metal can.*
■ *Place the open can on a stove, and heat until steam appears. Allow it to steam for 5 minutes.*
■ *Remove the can from the heat, and immediately put the cap on very tightly.*
■ *Wait!*

Results The can, like the plastic jug, will collapse. There will be much more noise as the metal starts to bend. The can can never be straightened, so use one that you plan to throw away.

16. Reversible Balloon

■ *Soak a piece of steel wool, about the size of an egg, in vinegar for 3 to 4 minutes.*
■ *Pull the steel wool into threads; use a pencil to push them into a glass bottle that has a small screw-top opening (or use a soda bottle).*
■ *Wet your fingers and allow 5 to 6 drops of water to fall into the bottle.*
■ *Attach a balloon to the mouth of the bottle.*
■ *Wait! It may take up to 24 hours for this experiment to be completed.*

Results The balloon will turn inside out and inflate inside the bottle.

Why? The vinegar removed the protective surface from the steel wool so that the iron in the steel would rust. The water helps it to rust faster. Iron combines with oxygen inside the bottle to form the solid iron oxide, rust. The rust takes up less space than the oxygen gas did, so the pressure inside the bottle is reduced. The air pressure outside the bottle pushes the balloon inside.

Try This Show the bottle with the balloon inflated inside it to a friend, and challenge him or her to produce a bottle with a balloon inflated inside. Give him or her all the materials that are visible— steel wool, soda bottle, and a balloon.

Results Your friend will unsuccessfully try to blow the balloon up inside the bottle; this is usually the result.

17. Rising Water

■ *Color 1 quart of water with your choice of food coloring.*
■ *Pour the colored water into a clear glass baking dish.*
■ *Light a candle, and stand it in the water. A short round candle works well, but you can use a slender candle and stand it up in the water by using modeling clay.*
■ *Slowly lower a dry, empty jar over the candle so that as the mouth of the jar touches the surface of the water, air does not bubble out from inside the jar.*
■ *Allow the jar to stand in the dish of water until the flame goes out.*

Results The water will start to rise in the jar.

Why? As the candle burns, the oxygen part of the air in the jar combines with the carbon from the melted candle wax. The product of this reaction is carbon dioxide gas. Oxygen molecules are lighter and faster than are carbon dioxide molecules. The carbon dioxide gas does not exert as much pressure as did the oxygen, so the air outside pushes the colored water into the jar.

Try This Use different sized jars, and observe the height of the water that is pushed inside by the air.

18. Heavy Air

■ *Lay a wooden yardstick on a table with about 7 inches extending past the edge. Since you are going to break the stick, use an inexpensive one and select one that is as thin as possible.*
■ *Separate 6 sheets of newspaper.*
■ *Lay the sheets individually and in different directions across the section of the stick lying on the table. Smooth each sheet of paper before adding another one.*
■ *Be sure the entire stick is covered with several layers of paper. Smooth each paper so as to remove any air under it before adding the next sheet.*
■ *Stand so that the exposed length of stick is in front of you and the hand you will strike the stick with is on the side away from the table.*
■ *Wear a glove to protect your hand.*
■ *Raise your hand and <u>quickly and forcefully</u> hit the end of the stick with the edge of your hand.*

Results The stick will break at the edge of the table.

Why? The air pushing on the paper causes the stick to be held in place as you hit the end, and the resistance of the table causes the stick to break.

Problems? Any possible problem with this experiment is probably due to your stopping short just before striking the stick and not hitting it with a quick stroke, or the paper has air pockets under it.

29

19. **Vooomm!!! Box**

- *Cut the top out of an empty salt box.*
- *Cut the neck from a 14 inch round rubber balloon. Stretch the rounded end over the opening in the salt box.*
- *Cut a hole about the size of a nickel in the center of the box's bottom.*
- *Light a candle, and set it in a saucer.*
- *Aim the hole in the box at the flame.*
- *Thump the stretched rubber with your finger.*

Results The candle flame will be blown out.

Why? Hitting the rubber compresses the air inside the box and forces it out the hole. This air is pushed forward with great force; its high speed when it hits the candle flame extinguishes it.

20. **Make Your Own Smog**

- *Rinse a clear glass, small-mouthed bottle with water. You want the inside of the bottle to be wet.*
- *Ignite a match, then blow it out. Collect a small amount of smoke in the bottle.*
- *Immediately insert a plastic aquarium air tube about 1 foot in*

30

length. Use modeling clay to seal the tube in the mouth of the bottle.

■ *Blow into the tube, then release.*

Results The contents of the bottle looks very smoggy, then clears up.

Why? An increase in pressure causes the temperature of a gas to increase. Blowing on the tube increases the pressure on the air inside the bottle and as a result, the temperature inside the bottle increases. Because of the increase in temperature, the drops of liquid water evaporate and form water vapor. The smoke particles inside the bottle stick to the large molecules of water vapor that are moving through the air in the bottle, producing smog. When you stop blowing and release the tube, there is a drastic reduction in the pressure inside the bottle, causing the water vapor to *condense* (change from gas to liquid). The liquid drops of water stick to the sides of the bottle, and there are no longer any large molecules in the air for the smoke particles to cling to, so the air clears up. There are always dust particles in the air, but smog can only be produced when there are enough dust particles and large molecules (like water vapor molecules) for the dust to stick to.

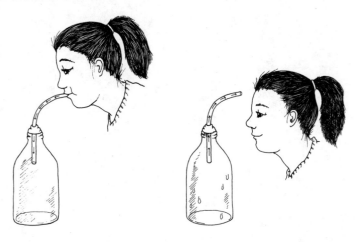

3

Bernoulli's Principle

3

Bernoulli's Principle

21. In Not Out

■ *Inflate 2 round balloons, and tie their ends.*
■ *Attach strings to the end of each balloon, leaving about 12 inches of free string.*
■ *Tape the free ends of the string to the edge of anything that will allow the balloons to swing freely about 6 inches apart.*
■ *Blow so that your breathe is directed in a straight line between the balloons.*

Results The balloons move toward each other.

Why? Daniel Bernoulli (1700–1782) observed that the faster air moves, the less pressure it exerts on objects above, below, to the right, or to the left. This is known as Bernoulli's principle. Another way of stating this principle is that the faster the gas moves, the more pressure exerted on objects perpendicular to the direction of the air's motion decreases. The air moving between the balloons exerts very little pressure on them. That is, the air flow directed between the two balloons has very little sideways pressure, and the air pressure on the outer sides of the balloons pushes them together.

22. Why Airplanes Fly

- *Fold a sheet of notebook paper in half, lengthwise.*
- *Open the sheet.*
- *Fold the top corners toward the center until they touch.*
- *Bring the folded edges toward the center until they touch, then crease them.*
- *Again bring the folded edges toward the center and crease.*
- *Fold backward, creasing the plane lengthwise.*
- *Raise the wings, hold the body, and throw.*

Results The paper plane sails through the air.

Why? The paper plane glides through the air because the air pushes up on the underside of the wing with greater force than it pushes down on the top side. This difference in air pressure exists because the shape of the wing causes the air to flow faster over the wing. Thus, this faster-moving air does not push down as much as the slower-moving air on the underside pushes up. Bernoulli's principle explains this force on the underside of the wing.

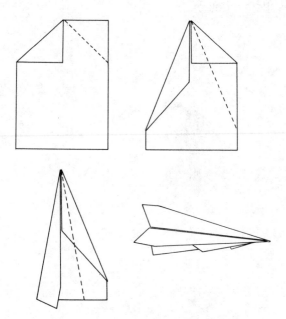

23. Paper Lift

■ *Cut a 1-inch strip from across a piece of notebook paper.*
■ *Hold one end of the paper against your chin, just below your bottom lip.*
■ *Blow across the top of the paper.*

Results The paper will lift toward the stream of air and flutter as long as you continue to blow.

Why? The paper lifts because the moving air across the top creates an area of low pressure. The air below it has a higher pressure and pushes it up. The fluttering is caused by the paper hitting the stream of air and being pushed down, only to be lifted again by the air below it.

24. Super Breath

■ *Cut a circle from an index card just slightly larger around than the end of a thread spool.*
■ *Stick a straight pin through the center of the circle.*
■ *Place the paper over the end of the spool with the pin inside the hole.*
■ *Turn the spool upside down, holding the paper against the spool.*

37

■ *Blow through the top end of the spool. You can stop holding the paper when you start blowing through the spool.*

Results The paper is not blown off, nor does it fall. It stays on the bottom of the spool.

Why? The air stream going through the spool escapes between the spool and the paper. This produces a low-pressure area between the paper and the spool. The air pressure on the underside of the paper is greater and holds the paper in place.

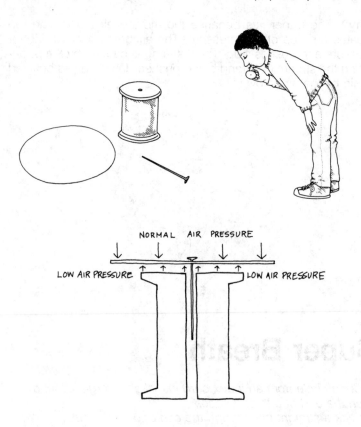

NORMAL AIR PRESSURE

LOW AIR PRESSURE   LOW AIR PRESSURE

4
Heat

25. Smoking Chimney

- *Cut the bottom from an empty oatmeal or cornmeal box.*
- *Cut a three-inch square from the edge of one end.*
- *Line the inside of the box with aluminum foil.*
- *Stand the lined box over a burning candle, with the cut-away section at the bottom.*
- *Have an adult light a cigarette and place it in an ash tray.*
- *Position the smoking cigarette near but not inside the opening in the box.*

Results The smoke will stream into the opening and up to the top of this box chimney.

Why? As the air inside the chimney heats up, the gas molecules move farther apart. This makes the air lighter, and it rises toward the top. As the warm air rises, the cooler air in the room rushes toward the opening to take its place. As it moves toward the door, the cooler air sweeps the cigarette smoke into the chimney with it. The smoke is carried out the top by the rising warm air. The movement of warm and cold air causes convection currents.

Watch for This The next time a person with a smoking cigarette is sitting near a burning lamp, watch the movement of the smoke up into and out the top of the lamp shade.

26. Spinning Spiral

- *Cut a 3-inch-diameter spiral from a sheet of notebook paper.*
- *Use the oatmeal box made for the "Smoking Chimney" experiment, or make a tube of aluminum foil to fit around a standing candle.*
- *Balance the center of the spiral on the point of a pencil. Be sure that you do not stick the point through the paper. This would restrict the movement of the paper spiral. This part takes practice, so don't give up if the spiral falls off the point a few times.*
- *Hold the spiral over the aluminum-foil or oatmeal-box chimney.*

Results The spiral whirls around.

Why? The heat from the flame heats the air above it, and the air molecules speed up and move farther apart. With the molecules farther apart, the air is lighter and moves upward. Cooler air from the room rushes in to take the place of the rising warm air. There will be a continuous movement of air, and this motion causes the paper spiral to rotate.

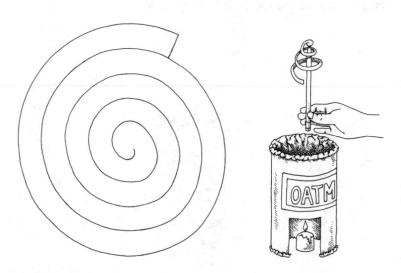

27. One Will, One Won't

■ *Fill a soda bottle to overflowing with hot, colored water. Food coloring can be used, and the water should be as dark as possible.*

■ *Fill another bottle with ice-cold water. (Allow ice cubes to sit in the water before pouring it into the bottle.)*

■ *Place a small piece of wax paper over the mouth of the cold-water bottle.*

■ *Hold the wax paper tightly so the water does not spill while you place the bottle upside down on top of the bottle containing hot water.*

■ *Be sure the mouths of the bottles line up, and then pull the wax paper out.*

Results The colored water will start to rise. In a short time the color in both bottles will be the same.

■ *Prepare the two bottles as in the first two steps.*

■ *Place the piece of wax paper over the mouth of the bottle containing hot water.*

■ *Turn the hot-water bottle over, and position it on top of the bottle containing the cold water.*

■ *Remove the wax paper.*

Results Little or nothing will happen. The small amount of colored water that moves down due to the motion of moving the bottles will rise.

Why? Heating water causes the molecules to move apart, and this makes the hot water lighter than the cold water. The lighter hot water rises, and the heavier cold water sinks. When the colored hot water is on top, it remains in the top bottle, and there is no mixing of the water.

43

28. What Is Needed to Cause a Fire?

Oxygen
■ *Place a candle on a table, and light it. The short, scented candles stand up well and are a good height for this experiment.*
■ *Use a quart-size, <u>glass</u> jar with a large mouth. A mayonnaise jar will work well.*
■ *Turn the jar over the burning candle.*
■ *Wait!*

Results The candle burns for a few seconds and then goes out.

Why? The air inside the jar contains many gases, and oxygen is just one of them. Oxygen is a requirement for the candle to burn, and when the oxygen is used up, the candle goes out.

Try This Use a gallon jar with a wide mouth. The time the candle burns will be much longer because of the greater amounts of oxygen in the jar. You can time the burning in different jars and compare them.

Fuel
■ *Use a taper candle.*
■ *Light the candle, and allow it to burn until there is a small pool of melted wax around the wick. Notice that the wick never burns all the way to the surface of the candle. The wax pool keeps it too cool to burn in this area.*
■ *Use metal tweezers to pinch the base of the wick. Be careful not to pinch the wick in half. Try to squeeze the part that is not charred.*

Results The flame will go out if you hold the wick long enough.

Why? The wick acts like a sponge that soaks up the melted wax. When this fuel supply is cut off, the flame goes out.

Kindling Temperature

■ *Use a metal strainer that has very small openings. It is important not to have large holes or tears in the metal.*
■ *Tear pieces of notebook paper or newspaper, and place them in the metal strainer.*
■ *Light a candle, and hold the flame under the strainer. A taper candle will give a larger flame, but any candle will do.*

Results The paper smokes, but you will find that it doesn't burn. The flame never goes through the strainer openings.

Why? The metal absorbs the heat from the flame, and the paper never reaches its *kindling temperature.* This refers to the heat needed to cause a substance to catch fire. The flame you see actually consists of small carbon particles that are not visible when cold; they glow when heated, producing the yellow light you recognize as a flame. When the flame hits the metal, the carbon particles stick to the metal, and the ones passing through are not hot enough to glow. Some of the carbon passing through the openings in the strainer will stick on the paper, but you will find very little of this.

For burning to occur there must be enough oxygen, a fuel, and enough heat for the fuel to reach the kindling temperature.

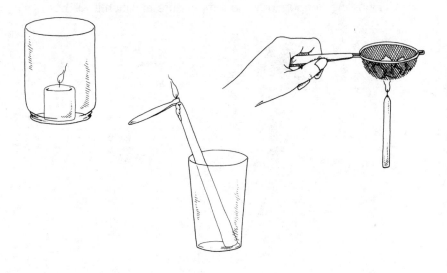

29. Boil Water in a Paper Cup

■ *Stick a meat fork through a paper muffin baking cup.*
■ *Pour in only enough water to cover the bottom of the cup with a thin layer. It is very important to cover the entire surface of the bottom of the paper cup.*
■ *Light a candle.*
■ *To secure the candle in an upright position, pour a few drops of melted wax in the kitchen sink, and stick the bottom of the candle in it.*
■ *Hold the cup above the flame. The tip of the candle flame must touch only the bottom of the cup. This area has water above it and will not catch fire. The sides of the paper cup can burn, so do not touch the candle flame to these areas.*
■ *Heat until the water boils.*

Results The water boils, but the paper doesn't burn.

Why? The water in the paper cup absorbs most of the heat given off by the candle flame. The water gets hotter, but the paper does not. The paper never gets hot enough to burn. It never reaches its kindling temperature, the temperature at which it will burn.

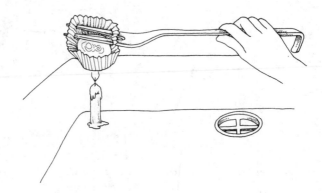

30. Golden Sparkles

■ Wrap a piece of steel wool around the prongs of a fork.
■ The steel wool must not contain soap. You will find steel wool balls that contain no soap in the same area of the grocery as the steel wool soap pads.
■ Separate some of the threads of the steel wool so that they hang down.
■ Tear off 2 sheets of aluminum foil, each about 2 feet long.
■ Lay the foil on a table to make a protective covering.
■ Place a candle in the center of the foil, and light it.
■ Hold the fork above the candle so that the steel threads hang into the flame.

Results Golden sparkles!

Why? The steel thread is so thin that most of the steel molecules are touched by oxygen molecules in the air. This makes it easy to burn. The heat from the candle is enough to cause the steel to reach its kindling temperature.

31. Paper That Won't Burn

- *Tear off 1 sheet of paper toweling.*
- *Place a quarter in the center of the paper.*
- *Pull the paper around the coin as tightly as possible without tearing it.*
- *Ask an adult to light a cigarette for this experiment.*
- *Press the burning end of the cigarette against the paper that is stretched over the quarter.*

Results The paper does not burn!

Why? The coin under the paper absorbs the heat from the lighted cigarette. The paper does not catch fire because it never gets hot enough to reach its kindling temperature.

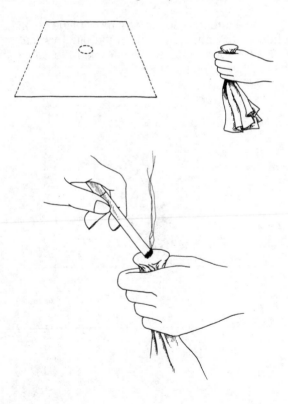

48

32. Catch an Ice Cube

- *Use cotton kite string.*
- *Cut 12 inches of string, and fray one end. (To fray means to cause the end to unravel so that small pieces stick out.)*
- *Lay an ice cube on a table.*
- *Place the frayed end of the string on top of the ice, and rub the string so that it lies flat against the ice.*
- *Sprinkle salt over the string where it touches the ice.*
- *Wait for about 1 minute.*
- *Gently pick up the string.*

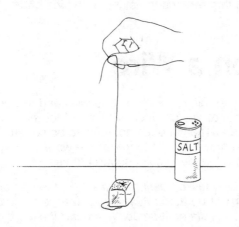

Results The ice cube will be stuck so firmly to the string that you can pick it up by raising the string.

Problems That Might Arise If it doesn't work, try again with another ice cube and a clean piece of string. You need to use a cotton string. Let the ice sit for a second until you see it slightly melting before you place the string on top.

Why? When salt dissolves in water, heat is removed from the water and the resulting salty water is colder than the unsalted water. Water without salt will freeze at zero degrees Celsius and salty water freezes a few degrees below zero, depending on the amount of salt that is dissolved in the water (the more salt that is dissolved, the lower the freezing point of the resulting salt water).

49

When the ice on the surface of the ice cube melts, the water formed is near the freezing point of zero for water. Dissolving salt in this cold water produces a salty water at a temperature below zero.

When pressed against the ice cube, the frayed ends of the cotton string absorb some of the water on the surface of the ice. Remember that this water will freeze at zero degrees Celsius. Adding the salt results in a salt solution with a temperature below zero degrees Celsius. The unsalted water in the string is surrounded by a subzero temperature and thus is cooled enough to freeze. Note that when the salt is added, some of it dissolves in the water in the string; but there is enough unsalted water in the string to freeze, and this secures the string to the ice cube.

33. Ice on a Wire

Before starting this experiment, you will need to freeze water in a container that will produce a block of ice no smaller than 4 inches square. The ice does not have to be square in shape—a plastic bowl may be used to freeze the water. It may take a day or two before the ice is solid enough to use.

■ *Remove the plastic coating from a 2-foot length of 18-gauge wire. You might be able to use larger wire, but it is difficult to work with and will not fit smoothly around the ice block. Smaller wire could break.*
■ *Fill 2 plastic gallon jugs with water, and put a cap on each.*
■ *Twist the ends of the wire several times around the handles of the water jugs.*
■ *Remove the ice from the plastic dish, and sit it on the corner of a table. If you have a table outside, it would be best to use it, as the ice will melt. If not, place a towel under the ice and another one on the floor to absorb the water.*
■ *Get help in holding the gallon water jugs, and place the wire connecting them across the center of the ice. Do not allow the wire to be supported by the edge of the table. The water jugs are to hang freely, with only the wire across the ice supporting them.*
■ *Allow the jugs to hang for several hours. Check them periodically.*

Results You will observe the wire cutting through the ice block. The ice refreezes as the wire passes through.

Why? The weight of the water jugs causes the wire to press against the ice. Pressure will cause the ice to melt, but the water formed is so cold that as soon as the wire moves down the water refreezes. Melting caused by pressure is called *regelation*.

Ice skaters are actually skating on water. The blades of the skates apply pressure to the ice and cause it to melt, just as your wire did. The continuous pressure of the skate blades makes a wet, slippery surface to skate on. When the ice is very cold, the pressure of the skate blades is not great enough to melt the ice, and it is difficult to skate.

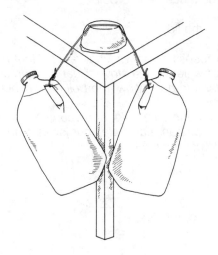

34. Boil Water with Ice

You will need an adult to help you with this experiment.

■ *Use a junior-size baby food jar with a tight-fitting lid.*
■ *Fill the jar half full of water.*
■ *Place the open jar in a microwave oven and heat. (The efficiency of different brands of microwave ovens makes the time vary.) When the water in the jar starts to boil, allow it to boil for 1 minute.*

■ *Open the microwave, and immediately place the lid on the jar.*
■ *Use care to remove the hot jar from the microwave, and after placing it on a table, tighten the lid.*
■ *As soon as the water stops boiling inside the jar, you can place an ice cube on top of the lid.*

Results The water in the jar starts to boil.
■ *Remove the ice, and dry the lid.*

Results The water stops boiling. If you place the ice cube on the lid again, the water will start to boil again; removal of the ice stops the boiling, etc. The water will continue to boil and stop until it loses too much heat to boil.

Do not remove the lid.

■ *When the process stops working, reactivate it by placing the jar in a pan of water on the stove. Heat the pan until the water in the pan starts to boil. Allow it to boil for 2 minutes, and then remove the jar.*
■ *Place an ice cube on the lid, and you will see the boiling action again. You can reactivate your jar as many times as you wish if you do not open the lid.*

Why?
■ *What is boiling?* When water or any liquid is heated in a pan, tiny bubbles of gas start collecting on the bottom of the pan. If it is water that is being heated, the bubbles are water vapor. Even though the bubbles form around the bottom and sides of the pan, the water has not reached its boiling point until the bubbles have enough energy to push up through the surface of the water. If the pan is uncovered, the air in the room will

push down on the surface of the water and keep the bubbles from escaping. Heat is a form of energy, and as the bubbles of water vapor are heated, they become more energized and move faster. They hit the undersurface of the water with greater force and break through when their force equals the force of the air pushing down on the water's surface. The boiling point of water is the temperature at which the bubbles of water vapor are energized enough to push up through the water's surface.

The boiling point can be changed by changing the pressure on the water's surface. If the pressure of the gas above the water's surface equals that of a pressure cooker, the water has a higher boiling point. This increase in pressure on the water's surface keeps the gas bubbles from escaping until they reach a higher temperature. A decrease in pressure on the water's surface allows the bubbles of water vapor to push up through the surface at a lower temperature, because there is less force keeping them in.

■ *What happened in the jar?* Heating the water in the open jar causes the air molecules above the water to heat up. Hot air rises: Thus boiling the water in the open jar causes most of the air above the water to heat up and leave the jar. When the jar is capped, there is liquid water, water vapor, and a small amount of air trapped in the jar.

Cooling the jar with ice makes the water vapor condense. Condensation is a change from a gas to a liquid. Since most of the air has been removed from the jar, there are very few air molecules to push down on the surface of the water. This decrease in pressure on the surface of the water allows the bubbles of water vapor inside the water to escape up through the water's surface more easily than if the surface pressure were greater. This decrease in pressure on the water's surface decreases the boiling point of the water.

Removal of the ice keeps the vapor produced by the boiling water in gas form. These molecules of water vapor above the water join the few air molecules; together they push much harder on the surface of the water and prevent additional bubbles of water vapor from escaping. Thus the boiling action stops.

5
Light
Refraction

35. Where's the Penny?

- *Use a cup or container that you cannot see through.*
- *Dry the inside bottom of the cup.*
- *Place a piece of modeling clay in the center of the cup.*
- *Stick a penny into the clay. This keeps the coin from moving around. The clay and the penny should stay in the cup even if it is turned upside down.*
- *Place the cup near the edge of a table.*
- *Stand so that you can see the penny in the cup.*
- *Slowly step backwards until the penny is just barely out of view. You should be able to see the penny if you lean forward a little.*
- *Stand straight in this position so that the penny is out of sight. Have another person pour water into the cup.*

Results The penny moves! No, the penny is still in the same place, but to your eye it appears to have moved.

Why? Look at the drawing of the person looking at the penny. Notice the dashed lines. They show the light comes straight up from the penny, bends when it hits the surface of the water, and travels to the eye. The light hits the eye, and the image of the penny is seen. Since the brain is programmed to respond to light that travels in a straight line, it tells you that the penny moved to a spot close to the edge of the cup. The bending of light is called *refraction*.

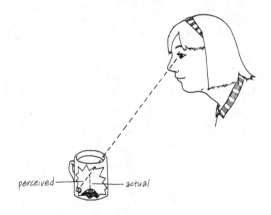

perceived ——— actual

36. Broken Pencil

- *Fill a clear drinking glass half full of water.*
- *Place a pencil in the glass.*
- *Stand so that you can see the top and side of the water.*

Results The pencil appears to be broken.

Why? You see the pencil because light bounces off of it to your eyes. The light from the side of the pencil in the air comes from a different direction than the light from the pencil parts under the water; this is because the light coming from the parts below the water bends when it hits the water's surface. This makes the pencil appear to be broken.

6
Lenses

37. Topsy-Turvy

- *Fill a junior-size baby food jar to overflowing with water.*
- *Place the lid on very tightly. The jar must not leak.*
- *Use ½-inch capital letters to write the word CHICK at the top of a sheet of white paper and the word TURK at the bottom of the same sheet of paper.*
- *Lay the water-filled jar sideways over the word CHICK.*

Results There is some magnification of the word when the bottle is lying on top of the page with the word under it.

- *Move the jar up and away from the paper until you can see the letters clearly through the jar.*
- *Repeat the fourth and fifth steps using the work TURK.*

Results The letters in the word TURK are enlarged and upside down, but the enlarged letters in the word CHICK appear to be upright.

Why? Actually, all the letters are turned over. The letters in CHICK are symmetrical letters and look the same even when upside down.

Try This Place a mark over each word, and again look at the words through the jar of water.

Results You can now tell that all the letters turn over in the image that you see through the water lens.

38. Water-Drop Microscope

- *Cut a piece of 20-gauge wire 6 inches long. Any small, bendable wire will work.*
- *If the wire is small enough, it is not necessary to remove the protective outer coating.*
- *Twist one end of the wire around a pencil to make a round loop.*
- *Remove the loop from the pencil, and make one more twist to decrease the size of the loop.*
- *Fill a deep bowl with water.*
- *Dip the loop into the water with the open loop pointing up.*
- *Lift the loop carefully out of the water. You are trying to get a large rounded drop of water to stay in the hole of your wire loop.*
- *Hold the water drop over a sheet of newspaper.*
- *Look through the water drop at the letters on the page. You may have to move the loop up and down a couple of times to find the position that makes the letters clearest.*

Results The letters will be enlarged.

Why? The water drop, if large enough, is rounded and acts like a double convex lens. It will most likely be rounded on both sides of the loop, as is a magnifying lens. If the letters do not look larger, try again to pick up a fat water drop.

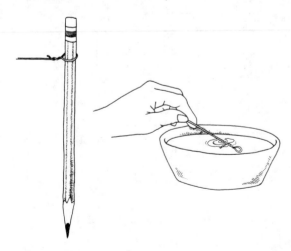

Try This Make a concave lens with your wire loop. There are two ways to make this lens, and you can try both to see which works best for you.

■ *Lift a fat water drop on the loop from the bowl of water, and then very carefully turn the loop completely over. The movement causes the water drop to sink, and it usually forms a caved-in drop. (Concave lenses cave in.) You may have to practice this to get a good lens.*
■ *Alternatively, fill the loop with water as you did in making a rounded convex lens. Very carefully touch the bottom of the water drop with the tip of your little finger. You are removing a very small amount of water from the drop, and again the drop will cave in to form a concave lens. Look through the concave water drop at letters on the newspaper sheet.*

Results The letters look smaller.

Why? Looking through a convex lens magnifies the object. A concave lens reduces the size of the object you are looking at. A single concave lens is formed in this experiment. Only one side is caved in. A double concave lens would reduce the image size even more. Because it is curved outward on the bottom, the concave lens is called a convexo-concave lens.

39. Tasty Magnifying Lens

■ *Cut a magnifying-lens holder from stiff paper.*
■ *Cut a lens of suitable size from a sheet of clear, see-through plastic to cover the opening in the lens holder.*
■ *Tape the plastic lens over one of the openings, fold the holder on the dotted line, and glue or tape the two sides together.*
■ *Use an eyedropper to place a large drop of colorless corn syrup in the center of the plastic lens.*
■ *Look through the syrup drop at letters on a newspaper sheet. You will have to move the holder up and down to focus the letters.*

- *Experiment with different amounts of syrup. Try with the syrup in a high mound, and then spread the drop out.*
- *Take the eyedropper and squeeze a bubble of air into the syrup drop. Look at the newsprint through the air bubble.*

Results The magnification increases where the syrup is thick and round.

Why? The thick round drops are like any convex lens. The thicker and rounder the lens, the more the image is enlarged.

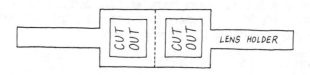

40. Pinhole Glasses

- *Cut a 2½-inch-diameter circle from stiff cardboard.*
- *Use a straight pin to make a small hole in the center of the circle. You want the hole just large enough to see through. Usually, you need to cut off the threads of paper sticking over the hole in order to see through it. Move the pin around to hollow out the hole; this will also help to push the hanging paper threads out of the way. Be careful; if the hole is too large, you will not be able to see a clear object.*
- *Shut one eye, and hold a newspaper sheet close enough to your eye so that the print is blurred. It helps to be in a well-lighted area.*
- *Put the cardboard between your eye and the newspaper, and look through the tiny hole in the cardboard with your open eye.*

Results You can read only a word, or maybe a couple of letters, at a time, but they are much clearer with the paper monacle than without it.

Why? As the newsprint is brought closer to the eye, more light from the paper reaches the eye. Overlapping of the light rays causes the image to be blurred. As you look through the tiny hole in the paper monacle, much of the excess light is shut out, and the letters are clear again.

Try This Look through your paper monacle at a burning light bulb. (Do not look directly at the bulb without the paper monacle.) Look at the area on top where there is writing.

Results The monacle allows you to read the words clearly.

Try This Have a person who wears corrective eyeglasses to remove them and, closing one eye, look through the paper monacle with the open eye at objects in the room. Provide a newspaper for him or her to read while looking through the monacle.

Results The person will be surprised that objects in the room can be clearly seen, or at least more clearly seen than without the monacle, and that the newspaper can be read.

Why? A person who needs corrective lenses can achieve better vision by squinting. This causes the lens of the eye to change shape. The lens also has to change shape in order to see through the monacle hole. This new shape improves the vision.

41. Camera Obscura

- *Line the insides of an empty coffee can with black construction paper.*
- *Use a nail to punch a small hole in the closed end of the can and through the black paper. Do not make the hole too large, as with an extra-large hole the image produced will be blurry.*
- *Cover the open end of the can with wax paper, and secure with a rubber band. The wax paper needs to be as smooth and wrinkle-free as possible.*
- *Wrap a piece of black construction paper around the outside of the can so that it extends about 3 inches past the end covered with wax paper. This is to provide a darkened area around the wax-paper screen.*
- *In a dark room hold the can so the wax-paper end faces you and the end with the hole is pointed toward an open window.*
- *Move the can back and forth from your eyes until you find the position that brings in the clearest image on the screen.*

Results On the wax-paper screen you will see the image of the window and objects outside. They are all upside down. The images on the retina of your eye are also upside down. Your brain sends you the message that things are right side up, as they really are.

Why? Light travels in a straight line. If the hole in the can is small enough, there is no overlapping of the light coming from the object, and it makes a clear image on the screen. Study the diagram to see how the light travels through the lens (hole in the can) to form the upside-down image on the wax-paper screen. The same thing happens as light goes through the lens of your eye to the retina.

Try This Have someone stand in front of the open window and wave his or her hands overhead as you look at him or her with your camera.

Results The person will appear to be standing on his or her head and waving his or her arms.

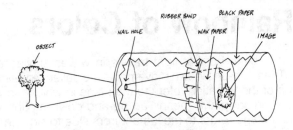

42. A Living Picture

■ *Hold a magnifying lens so that the light from a window passes through it. (This experiment works best when the sun is very bright and the room is dark.)*
■ *Move a sheet of white paper back and forth from the magnifyer until a clear image of the window and objects outside appears. (Hold the paper on the opposite side of the magnifyer from the window.)*

Results You can see a picture of the window and objects outside on the paper screen. The objects and window are upside down.

Why? Images formed by the convex lens in the magnifyer are always real and inverted. Real images are ones that can be projected onto a screen. The "Why?" section of the *Camera Obscura* experiment explains how light travels through a convex lens.

Try This Have someone walk outside near the window while you use the magnifyer to project the image onto the paper screen.

Results You will see the person walking upside down on your paper screen.

43. A Rainbow of Colors

Facts It is believed that light travels in waves similar to water waves. These up-and-down movements are called vibrations. The direction of the vibration refers to the plane in which the wave is moving. Vertical plane means an up-and-down motion while a horizontal plane refers to movement from side to side. In a room the vertical plane would be from the ceiling to the floor and the horizontal plane is from one wall across the room to the opposite wall. Sunlight contains waves of light that *vibrate* (move up and down like water waves) vertically, horizontally, and at all angles in between.

■ *Take the lenses out of a pair of polarized sunglasses. If you are unable to take a pair apart, then use two pairs of polarized sunglasses. You just need to hold one lens over the other.*
■ *Hold one lens about 1 foot from your face, and point it toward a source of light (such as an open window).*
■ *Hold the second lens directly behind the first lens.*
■ *Rotate only one lens as you look through both lenses.*

Results The light is blocked out with every 90° rotation.

Why? The polarized lens acts as if it had vertical slits that allow only vertically vibrating waves of light to pass through. When the lens is rotated 90°, the "imaginary slits" would be in a horizontal direction and would allow only horizontally vibrating waves of light through. Note that polarized lenses only act as if they had slits, and that is why the words "imaginary slits" are used to describe what happens.

As you look through two lenses and rotate only one of them, you will see that both lenses allow light with a specific vibration to pass through. A point is reached where one lens allows a specific vibration through, but the second lens blocks it. When this happens, little or no light passes through the lenses and they look dark.

Try This
■ *Use stiff cardboard to make a paper frame 2 inches square with a 1½-inch square cut out of the center.*
■ *Use very cheap cellophane tape. You want the kind that will turn yellow with age. The poorer the quality of the tape, the better for use in this experiment.*

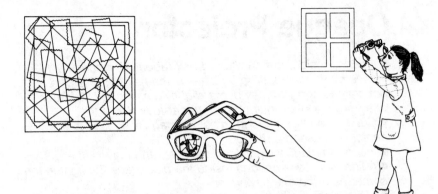

- *Cover the opening in the paper frame with strips of the cellophane tape. Overlap the pieces, and put them on the front and back so that you cover up the sticky sides of the tape. Lay the pieces in different directions as you overlap them.*
- *Continue to add tape until there are no open spaces.*
- *Hold one of the polarized lenses under the frame of tape.*
- *Hold the second polarized lens on top of the frame of tape.*
- *Look through the lenses and tape, holding one lens stationary while rotating the other lens.*

Results The tape takes on many different colors. If it doesn't, you have either used tape that is too good (usually called *invisible* tape) or the lenses are not polarized. To check the lenses for polarization, hold one lens in front of the other. Slowly rotate only one of the lenses as you look through them. They will turn very dark and shut out all or most of the light if they are polarized.

Why? Sunlight contains all the colors of the rainbow: red, orange, yellow, green, blue, indigo, and violet. These colors are all vibrating in the same direction as they pass through the first polarized lens. The tape does a strange thing to these vibrating colors; it seems to twist them slightly, so their vibrations move in different directions. They may have passed through the first lens moving in an up-and-down direction, but after passing through the tape, red may be horizontal, violet vertical, and the other colors at various angles in between. All of the colors pass through the tape, but as they hit the second polarized lens only those colors vibrating in the direction of the "imaginary slits" in the lens are allowed through. Therefore, you see the colors separately, and as the lens is rotated, the colors vibrating at different angles are allowed through.

44. Opaque Projector

■ Line the inside of a cardboard box (about 18 inches by 11 inches by 9 inches). (The 9 inches is the height.) This size is not critical, and you may use a slightly smaller box. A larger one will work but requires more paper to line it.
■ Cut a hole large enough to hold a magnifying lens in the center of one end of the box.
■ Insert the magnifying lens in the hole, and leave it there.
■ A light source is needed inside the box. Place the light in the corner near the magnifying lens.
■ Make a screen by taping a piece of white paper to the side of another box.
■ Position the screen in front of the magnifying lens.
■ Hold an open book inside the box with the pages facing the lens.
■ Move the book back and forth from the lens to produce a sharp picture on the screen. If moving the book does not produce a sharp picture, then change the position of the screen.
■ This experiment works best in a dark room.

Results You can see an upside-down image of the book on the paper screen.

Why? Light passes from the book through the convex lens of the magnifyer, and a real image is projected on the screen. See the *Camera Obscura* experiment for a diagram of how the light travels through a convex lens to form an upside-down image.

Try This Hold your hand inside the box, and wiggle your fingers.

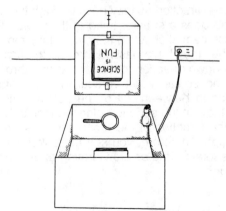

7
Reflection

45. Fire Under Water?

■ *Position the materials on a table in the order shown in the diagram: cardboard, candle, glass plate, glass of water.*
■ *Stand a piece of cardboard up in front of the candle with modeling clay. The cardboard must be tall enough to block your view of the burning candle. The height of the cardboard will depend on how far away you stand from the table (see the last step).*
■ *Stand a candle next.*
■ *Use a piece of modeling clay to stand a clear glass plate upright. This glass can be taken from a 5-inch by 7-inch picture frame. Be sure the glass is very clean.*
■ *Next, place a clear glass, three quarters full of water.*
■ *Light the candle.*
■ *Stand so that the cardboard blocks your view of the actual candle, but so that you can look through the glass plate and into the glass of water.*
■ *The burning candle seems to be burning under the water inside the glass. If this image does not appear, the distance between the candle and glass of water must be changed. Move these items until the image of the burning candle does appear in the glass of water. If the flame is above the water level, add more water to the glass or use a shorter candle.*

Remember that you are not to look at the burning candle, but through the glass plate and into the glass of water.

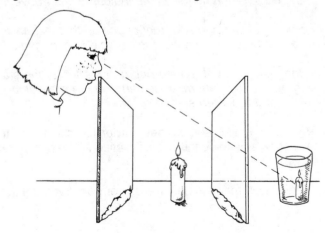

Why? The clear glass plate acts like a window and a mirror. You see the glass of water as you look through the glass plate, and at the same time you see the reflection of the burning candle on the glass plate.

Try This Empty the water from the glass, and replace it. Ask someone to stand so that the image of the burning candle appears to be in the empty glass. As he or she watches, carefully and slowly pour water into the glass until it is full.

Results The water slowly rises and should put the candle out but does not. It appears that the candle continues to burn under the water.

46. Star Race

■ Cut a block of wood large enough so you can glue a mirror to it that is at least 2 by 3 inches in size.
■ Glue the mirror to the wooden block so that the long edge of the mirror rests on the table when the block is standing. The block must be level, as the mirror must make a right angle with the table.
■ Draw a star pattern onto a sheet of paper.
■ Place the edge of the mirror on the indicated line above the star.
■ Hold a piece of stiff paper in front of you so that the star pattern cannot be seen. You want to be able to see the star in the mirror only.
■ Place the point of a pen on the paper at the top of the star.
■ Looking only at the pattern in the mirror, try to draw along the inside edge of the printed star.

Results You may find some sides harder to draw than others and may not be able to do this at all. Practice will help to increase your speed.

Why? The mirror image is backwards from what your mind tells you.

74

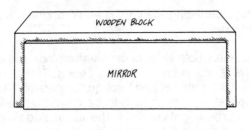

WOODEN BLOCK

MIRROR

PLACE MIRROR ON THIS LINE

Try This Challenge friends to draw the star, and time each person. You can beat them if you practice a few times before making the challenge.

Try This Looking only in the mirror, write your name so that it appears correctly in the mirror image.

47. Light Meter

■ Cut a 2½-inch by 5-inch by ½-inch block of paraffin in half to form two 2½-inch squares. (Paraffin can be purchased in block form in the canning department of your grocery store.)
■ Cut a piece of aluminum foil 2½-inches by 5-inches, and fold with the shiny side out to form a 2½-inch square.
■ Sandwich the folded foil between the two pieces of paraffin.
■ Hold the sandwich together with a rubber band.
■ Use a window or lamp for a light source.
■ Hold the sandwich so that you are looking at a side view, with aluminum foil separating the layers, and one of the paraffin sides is facing the light.
■ Turn the sandwich so that the dark side faces the light.

Results The paraffin facing the light is bright while the other paraffin piece is dark.

Results The side facing the light is again brighter than the one on the other side.

Why? Light enters both sides of the paraffin block and is reflected by the piece of aluminum foil in the center. The aluminum foil prevents the light from passing through the paraffin. The brighter side of the paraffin block is said to be more *illuminated*. Illuminated means to be bright with light. The illuminated side has more light that enters and is reflected by the aluminum foil than does the darker side. When both sides appear to have equal illumination, an equal amount of light is entering each side and being reflected by the aluminum foil.

Try This Use your meter to find areas of equal light intensity, those in which both sides of the meter look the same. You could also use your light meter to find the best lighted area in your house to read a good book.

48. Periscope

Instructions will be given to fit a specific size holder. If you choose to use different materials, the size of the mirrors may need to be changed.

■ *Open an empty wax-paper or aluminum-foil box.*

■ *Two mirrors that are 2¼-inches by 2¾-inches need to be positioned as shown in the diagram.*
■ *Hold the box up carefully so that the mirrors do not fall out, and check their positions. You may need to secure the mirrors loosely with clear tape to keep them in place while checking. You will be able to see the reflection of the objects behind you in the bottom mirror if they are in the correct positions. Adjust the mirrors if the reflection is not right.*
■ *When the mirrors are in the correct positions, use clear tape to secure them firmly to the box.*
■ *In the lid of the box cut a square 1½-inch by 1½-inch so that the opening will be directly over a mirror when the lid is closed. Do this over both mirrors.*
■ *Close the lid, and tape it.*
■ *Look through the bottom hole.*

Results You can see the objects behind you. They will be upside down in the reflection.

Why? Mirror images are always backwards. Left is right and right is left in the reflection. The image in the top mirror is backwards but not upside down. The image from the top mirror is reflected upside down to your eyes because the mirrors are at an angle to each other. The angle makes the bottom part of the image on the top mirror appear as the top of the image on the bottom mirror. The illustration shows how this happens.

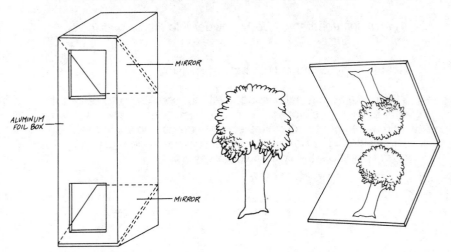

49. Why Outer Space Is Dark

■ *If you made an opaque projector (Experiment #44), remove the magnifying lens and use the box for this experiment. If not, follow these instructions to make a box to use. (a) Remove the top from a box that is about 18-inches wide by 11-inches long by 9-inches deep. These dimensions are not critical, but you need one about this size. (b) Cover the inside of the box with black construction paper. (c) Cut a hole in one end large enough for the light of a strong flashlight to shine through. A circle with a 3-inch-diameter will do.*

■ *Turn the box on its side with the open top facing out.*

■ *This experiment must be done in a very dark room.*

■ *Hold a flashlight outside the box as far from the hole as possible to enable you still to produce a circle of light on the inside of the box opposite the hole. (You are to shine the light in without having it reflect off the sides and illuminate the inside of the box.)*

Results The inside of the box will be slightly illuminated, but not much if you hold the light correctly. There will be a circle of light projected on the side of the box opposite the hole, but you will not be able to see the beam of light passing through the box.

■ *Have someone spray body powder into the box while you hold the light.*

Results The beam of light becomes visible.

Why? Just as in outer space, without something to reflect the light to your eyes, you do not see the light. The specks of powder reflect the light to your eyes so that you can now see the beam. The inside of the box will be more illuminated because the white powder reflects the light to your eyes.

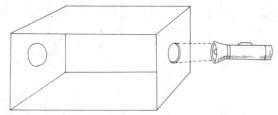

8
Color

50. Color Wheel

- ■ Use a large, flat button with two holes in the center.
- ■ Divide both sides of the button into three equal parts by drawing pencil lines toward the center.
- ■ Color one of the divisions blue, one red, and the third green. Do this on both sides.
- ■ Run a string 30 inches long through one of the holes and back through the other one.
- ■ Tie the ends of the string together, and move the button to the center of the string.
- ■ Turn the button around until the string is tightly twisted.
- ■ Pull outward on both ends of the string until it unwinds, then release the tension on the string ends so that it winds in the opposite direction. Pull—release—pull—release.

Results The button will quickly spin as you pull on the string and then turn in the opposite direction as the string rewinds. The colors blend together, and you should see the button as a grayish white color.

Why? Your eyes retain the separate image of each color for about 1/16 of a second after it has passed by. There is an overlapping of the colors in your mind. Your eyes actually blend the colors together.

If you could use a pure blue, red, and yellow to paint the button, the blending of the light from these primary colors would produce white light. It is difficult to get the exact shades of these colors that are needed to produce the white light. Grayish white is a good result for the experiment.

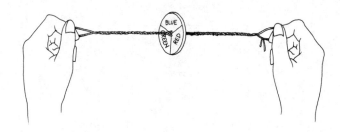

51.Make a Color

■ *Cut 2-inch squares from colored, see-through plastic folders. Colored cellophane does not work well.*
■ *You need red, blue, and yellow plastic squares.*
■ *Lay the plastic pieces so that they overlap as in the diagram.*

Results When colored materials like these plastic pieces or paint pigments are combined, the blending of the primary colors, blue, red, and yellow, will produce a black color. If the right shades of the colors are used, black will be the result, but usually the best result you can expect is a grayish black.

Remember that when blue, red, and yellow light rays are blended, the result is a white light. If the colors being blended are pigments, the color produced is black.

Try to produce these combinations with your plastic pieces:

COLOR COMBINATION	RESULTING COLOR
Red + Yellow	Orange
Red + Blue	Purple
Yellow + Blue	Green
Red + Blue + Yellow	Grayish Black to Black

Why? White light contains all the colors that are seen. The color of an object depends on the portion of white light that is reflected. A red dress looks red because all the colors in the white light hitting it are absorbed, except the red, which is reflected to your eye. Remember that you see objects because light is reflected to your eye. You see color because that part of the light is reflected to your eye.

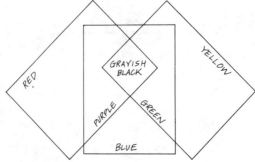

52. Secret Message

■ *Make an envelope: (a) Fold a piece of unlined white paper, and draw an envelope pattern on it, 3½-inches square; (b) Cut out a heart from one layer of the paper; (c) Fold, and tape the two sides together.*

■ *Use a black or blue pen to write* <u>Happy Valentine's Day</u> *across the center of the heart.*

■ *Use an* <u>orange felt tip pen</u> *to write your secret message inside the heart, near but not over the other writing. A yellow felt-tip pen will also work.*

■ *Cut a piece from a clear, red, see-through plastic folder to fit inside the paper envelope.*

■ *Insert the red plastic in the envelope.*

Results The orange writing disappears, and only the black or blue print is visible.

Why? The red plastic sends red light rays to your eyes. The orange tends to blend in with the colored plastic, since the color is usually not a true red but more of a reddish orange. Your eyes are not selective enough to distinguish the slight difference in the light coming from the plastic and the area of the orange writing. The color from the pen is thus camouflaged by the plastic.

CUT THIS SECTION
OUT OF
ONE LAYER
OF PAPER

HAPPY
VALENTINE'S
secret
message
in
orange
ink

53. Colored Sand

■ *Secure some fine sand by: (a) Using sand from the beach—sift out debris, and allow the sand to dry. (b) Obtaining the sand from a company that does sand blasting. (c) Obtaining the sand from a company that makes glass.*
■ *Place 1 cup of dry sand into a plastic bag.*
■ *Add 2 teaspoons of dry tempera paint (any color).*
■ *Hold the top of the bag, and shake until the color is uniform. Add more tempera paint if the color is not intense enough.*
■ *Set this colored sand aside, and continue to make bags of colored sand. (Keep these bags of sand to use in the* Sand Painting *experiment.)*
■ *If you have only the 3 primary colors of red, blue, and yellow, then follow the chart below to produce other colors. Experiment by adding different amounts of colored sand together to produce your own colors.*

COLOR COMBINATION	RESULTING COLOR
Yellow + Blue	Green
Yellow + Red	Orange
Red + Blue	Purple
Red + Blue + Yellow	Grayish Black

54. Sand Painting

■ *Clean and dry a small baby food jar.*
■ *You need bags of different colors of sand. (Instructions for making colored sand are given in the* Colored Sand *experiment.)*
■ *Use a teaspoon to pour sand around the inside edge of the baby food jar.*
■ *Alternate colors of sand to produce bands of colors.*
■ *Build up areas to produce mountains or other designs.*

■ *Make an effort to place the sand near the glass. This will leave an empty space in the center of the jar. Fill this cavity with white sand as you go. The white sand will not be seen from the outside when you finish.*
■ *Experiment with designs. Punch a pencil or small wire down the side of the glass into the sand to move the sand around.*
■ *Fill the jar completely to the top with sand.*
■ *Securely tighten the lid on the top. Never shake the jar!*

9
Magnets

55. Floating Magnets

■ *Purchase several round magnets with holes in their centers (available at most electronics stores).*
■ *Place a pencil through the center of at least 3 of the magnets.*
■ *Hold the pencil in an upright position.*

Results The magnets will float separately with the bottom magnet resting against your fingers. If any or all of the magnets stick together, you must remove one of them, turn it over, and replace it on the pencil. Continue to do this with the magnets until they float separately.

Why? The magnets have a magnetic force field around them. This force field flows out of the side of the magnet called the *north pole* and back into the other side, called the *south pole.* The north pole of one magnet attracts the south pole of another magnet. Like poles repel each other. North repels north, and south repels south. For the magnets to float separately on the pencil, like poles must be facing each other. There is enough force between the magnets to support their weight and thus they float.

Try This Push the magnets down toward your hand and then release them.

Results They spring apart.

Why? The repelling force between the magnets pushes them up.

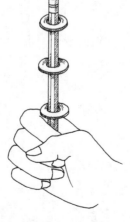

56. Swinging Compass

■ *Magnetize a large sewing needle by laying it on a magnet for 2 to 3 minutes. The needle will now be referred to as the magnet.*
■ *Tie a sewing thread to the center of the magnet, and hang the magnet inside a glass quart jar.*
■ *Lay a pencil across the jar opening, and tape the free end of the thread to it.*
■ *Allow the magnet to swing freely.*

Results The ends of the magnet point in the directions of north and south. The glass jar protects the magnet from air currents.

Why? The north pole of a magnet is actually the north-seeking pole. The earth is like a giant magnet, with a force field coming out of the south end and around to the north end, where it enters. The force field of your magnet, when it is allowed to swing freely, aligns itself with the magnetic field moving around the earth. The north pole of the magnet, when it is swinging freely, points to the magnetic north pole of the earth.

Try This Swing the magnet gently to try to point it in a different direction.

Results The magnet, when the energy of the swing is gone, will again point to the north.

57. Floating Compass

■ *Lay a large sewing needle on top of a bar magnet with the eye of the needle pointing to the south pole of the magnet. The needle must be made of a material that is attracted to the magnet for the experiment to work. If the needle will not stick to the magnet, try another brand of sewing needle.*
■ *Fill a cereal bowl with water.*
■ *Cut a sponge into a cube that is about ½-inch by ½-inch by ½-inch. You just want a small piece of sponge that will float on top of the water and hold the needle.*
■ *Place the sponge cube in the bowl of water.*
■ *After 5 minutes remove the needle from the magnet, and place it on top of the sponge piece in the water.*

Results The needle and the sponge swing around so that the tip of the needle points northward.

Why? Only materials that are attracted to a magnet can themselves become magnets. The force field of the bar magnet flowed through the needle, causing the direction of the electrons inside to change. The more electrons that spin in the same direction, the more magnetic is the material.

The force field of the needle is strong enough to move the floating sponge around. The needle will align itself with the earth's magnetic field, and the point of the needle will be pointing to the north.

Try This
■ *Instead of a sponge cut a small piece of cardboard to float the needle on.*

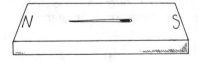

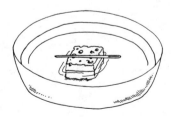

■ *Cut a drinking straw just a little longer than the needle. Lay the magnetic needle inside. You will need to mark the end of the straw with the point of the needle because this is the end that will be pointing north.*

Did You Know? This type of floating compass is similar to the one that Christopher Columbus used in making his ocean voyage.

58. Magnetized Boats

■ *Lay a sewing needle on a bar magnet for 5 minutes to magnetize it. One needle is needed for each boat that you plan to make.*
■ *Fill a large bowl with water.*
■ *Make one of the following boats:*

   <u>Cardboard Racer</u>: *Place a magnetized needle on a triangular-shaped piece of lightweight cardboard.*
   <u>Sponge Boat</u>: *Cut a cube from a piece of sponge, and insert the magnetized needle through the center.*
   <u>Cork Barge</u>: *Cut a cork, and insert a magnetized needle through the center.*
   <u>Straw Raft</u>: *Tie three straws together with sewing thread. Insert a magnetized needle in the center straw.*

■ *Position the magnetized needle in or on the boat.*
■ *Place the boat in the bowl of water.*
■ *Move a magnet around the outside of the bowl.*

Results The boat will follow the magnet. It moves across the water as you move the magnet around.

Why? The magnetic attraction for the needle is strong enough to pass through the glass bowl. This will work even if you do not magnetize the needle, but the force of attraction is greater if you do.

92

59. Flying Frog

- *Draw a frog on lightweight paper.*
- *Color the frog green, and add the eyes.*
- *Cut the frog design out.*
- *Tie a sewing thread to a small paper clip.*
- *Glue the paper clip to the back of the paper frog. Do not use an excessive amount of glue because you do not want to add extra weight to the frog.*
- *Lay a bar magnet or any type of strong magnet over the edge of a table.*
- *When the glue dries on the frog, hold it so that the paper clip sticks to the magnet.*
- *Slowly pull on the end of the thread to move the frog away from the magnet. The strength of the magnet will determine the distance you can move the frog.*
- *When you find the place where the frog will stay suspended in the air, tape the end of the thread to the floor.*
- *To add a touch of mystery cover the top of the magnet with a box or cloth.*

Results The frog appears to be leaping through the air and remains suspended there.

Why? The paper clip is attached to the paper frog, and the metal in the clip is attracted to the magnet. As long as the weight of the paper, glue, and thread are not too much, the magnet can pull them and overcome the force of gravity trying to pull them down.

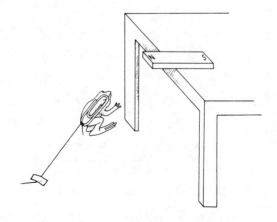

60. Magnetic Fishing Pole

■ *Use a 3/16-inch wooden dowel rod about 3 feet long for your fishing pole. (A small branch from a tree would work. You might even use a real fishing pole or rod.) Attach a 3-foot string to the end.*

■ *Instead of a hook, attach a round magnet with a hole in the center.*

■ *Make fish in one of these ways:*

Draw fish or cut out small fish pictures. Write a name on each fish. Fish out a name or names for volunteers to do jobs, answer questions, etc. Be sure to tape a paper clip to each one.

Write questions on small pieces of paper, and tape a small paper clip to the back of each slip of paper. Have someone fish for their own questions to answer.

■ *Each fish must have a paper clip taped to its side. Use small paper clips to minimize the weight of the fish.*

■ *Place the fish in a box, and dangle the magnetic hook in to catch them.*

Results The paper clips stick to the magnet, and one or more fish are pulled up.

Why? The paper clips are magnetically attracted to the magnet.

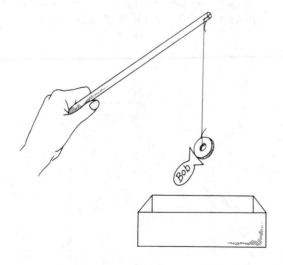

10
Electricity

61. How to Charge a Balloon

This experiment works best when the air temperature is cold and humidity is very low. It will work with a high temperature but not with a high humidity.

- *Blow up a balloon, and tie it. The size of the balloon is not important. Use one that you can easily hold in your hand.*
- *Rub the balloon against your hair, about 10 strokes.*

Results The balloon becomes electrically charged.

Why? All matter is made up of atoms that have a positive center with negatively charged electrons spinning around on the outside. Rubbing the balloon against your hair actually rubs electrons off of the hair and onto the balloon. This makes the balloon negatively charged and leaves the hair more positively charged.

Try This After rubbing the balloon on your hair, hold it close to, but not touching, the hair.

Results Your hair will stand on end.

Why? The negatively charged balloon is attracted to your positively charged hair. This force of attraction is strong enough to overcome the pull of gravity and lift the strands of hair. (Clean, dry hair works best. Oily hair can be too heavy to lift.)

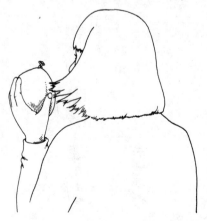

62. Defy Gravity

■ *Charge a long, slender balloon by rubbing it on your hair.*
■ *Stand on a table, and gently touch the longest side of the balloon to the ceiling.*
■ *Remove your hand from the balloon.*

Results The balloon will defy gravity and stick to the ceiling.

Why? The atoms in the ceiling, like all atoms, have a positive center with negative electrons spinning around. Bringing the negatively charged balloon near the ceiling causes the electrons in the ceiling to move away. This motion is due to the repulsion that like charges have for each other. The motion also produces an area on the ceiling near the balloon with a slightly positive charge. The negative balloon and the positively charged area on the ceiling are attracted to each other. This electrical force of attraction is strong enough to hold the balloon to the ceiling against the force of gravity, which pulls down on the balloon.

The balloon will stick to the ceiling for different periods of time on different days, depending on the humidity and air flow in the room. As air or water molecules hit the balloon, the excess electrons are knocked off.

Problems That Might Occur
■ If it did not stick:

  You may not have charged the balloon enough.
  The humidity may be too high; if so, try again on a drier day.

63. Spraying Water

■ *Charge an inflated balloon by rubbing it on your hair.*
■ *Turn on the water faucet in your kitchen sink so that the water falls in a small but continuous stream.*
■ *Hold the charged balloon near but not touching the water stream.*

Results The water will bend toward the balloon, and small streams of water will spray out.

Why? Each water molecule contains two hydrogen atoms and one oxygen atom. The molecule has a physical appearance similar to a round face with ears. Notice that one side is positively charged and the other negatively charged. The balloon picks up electrons as you rub it against your hair. The negative charge on the balloon attracts the positive, hydrogen side of the water molecule. The attraction between the negative and positive charges is strong enough to pull the whole water molecule toward the balloon.

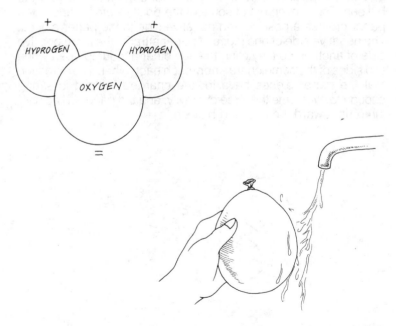

64. Dancing Papers

- *Tear 10 pieces of paper, making each about the size of the tip of your index finger.*
- *Place the paper pieces in a clear, plastic drinking tumbler.*
- *Cut a circle of paper to fit the top of the tumbler.*
- *Glue the paper circle to the top of the tumbler.*
- *Allow the glue to dry.*
- *Charge an inflated balloon by rubbing it on your hair.*
- *Hold the charged balloon close to, but not touching, the tumbler.*

Results The paper pieces will start to dance and jump around inside.

Why? All matter is made up of atoms that have a positive center with negatively charged electrons spinning around outside. Rubbing the balloon against your hair actually rubs electrons off the hair and onto the balloon. This makes the balloon negatively charged; when this charged balloon is brought near the paper pieces, the positive part of the paper tends to move toward the balloon. This movement of some of the positive atom parts in the paper creates a positive and negative side in the paper pieces. The negative side of one piece of paper attracts the more positive side of another, so they cling to each other and jump apart when two sides of the same charge approach each other. The attraction that the paper pieces have for the charged balloon is strong enough to overcome the force of gravity, and the paper will actually jump up toward the charged balloon.

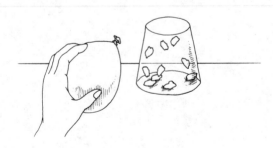

65. Charged Comb

■ *Charge a plastic comb by running it through your hair several times. This works best when your hair is clean and dry.*
■ *Tear a small section of lightweight paper into 10 pieces, each about the size of a dime.*
■ *Hold the charged comb near the paper pieces as they lie on a table.*

Results The paper pieces will jump to the comb, and some will cling to each other.

Why? Like all atoms, the atoms in the hair have a positive center with negative electrons spinning around outside. Running the plastic comb through the hair knocks off some of the electrons which stick to the comb and give it an excess number of negative charges. The negatively charged comb attracts the positive part of the atoms in the paper pieces. This attraction is strong enough to overcome the force of gravity, and the paper will actually jump up toward the charged comb. Because some of the positive parts of the atoms in the paper pieces move to one side as the negatively charged comb is brought near, the paper pieces have a positive and negative side. The negative side of one piece of paper attracts the more positive side of another, and they cling to each other.

65

66. Fluttering Butterfly

■ *Draw a butterfly onto a 3-inch-square piece of tissue or other lightweight paper.*
■ *Cut out the butterfly design.*
■ *Put a small amount of glue on the body, and glue it to a piece of cardboard. Do not glue the wing down.*
■ *Allow the glue to dry.*
■ *Crease the wings next to the body so they will bend up and down easily.*
■ *Charge an inflated balloon by rubbing it on your hair.*
■ *Hold the charged balloon near the wings, then move it away.*

Results The wings flutter as the balloon is brought near, then removed.

Why? All matter is made up of atoms that have a positive center with negatively charged electrons moving around outside. Electrons on the hair are knocked off and picked up by the balloon as it is rubbed against the hair. This makes the balloon have a negative charge. When this charged balloon is brought near the paper butterfly, the positive part of the paper tends to move toward the balloon. This movement of some of the positive atom parts in the paper causes the paper nearer the balloon to be more positively charged. The attraction of the now positive paper and the negatively charged balloon is strong enough to overcome the pull of gravity. When the balloon is moved away from the paper butterfly, the wings are pulled down by gravity.

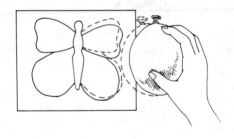

67. Are You Attractive?

■ *Tie a string to an inflated balloon. Leave about 18 inches of free string.*
■ *Tape the free end of the string to the edge of a table or anything that will allow the balloon to swing freely.*
■ *Charge the balloon by rubbing it on your hair.*
■ *Hold your hand near, but not touching, the balloon.*

Results The balloon is attracted to your hand and moves toward your hand.

Why? When you rub a balloon on your hair, electrons from your hair are picked up and the balloon becomes negatively charged. You are left with a slightly positive charge. There is a strong attraction between your hand and the balloon. When the balloon moves toward you, you witness this strong attraction between your hand and the balloon.

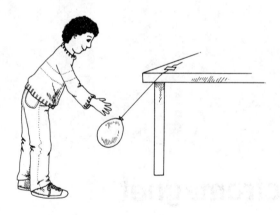

68. How a Flashlight Works

■ *Cut a 12-inch strip of aluminum foil about ½-inch wide.*
■ *Fold the foil in half lengthwise.*
■ *Wrap one end of the foil around the threads of a flashlight bulb.*
■ *Stand the negative end of a flashlight battery on the free end of the foil strip.*
■ *Hold the metal bottom of the bulb on top of the positive end of the battery.*

Results The bulb lights.

Why? Electrons flow from the battery through the bulb and back to the battery. The flow of electrons heats up the wire filament in the bulb. The hot filament wire causes the gas inside the bulb to glow.

69. Electromagnet

■ *Tape 2 size D batteries together with the positive end of one touching the negative end of the other.*
■ *Cut a strip of aluminum foil 2-inches wide and about 30-inches long.*
■ *Fold the foil lengthwise 3 times, which will result in the strip being ¼-inch wide by 30-inches long.*
■ *Wrap the foil around a long iron nail, leaving about 6 inches of free foil on each end.*

104

- *Stand the negative end of the batteries on one end of the foil.*
- *Hold the other free end of the foil on the positive end of the battery.*
- *Touch the nail tip to small metal objects like paper clips, tacks, or staples.*

Results The nail acts like a magnet and is powerful enough to pick up lightweight metal objects.

Why? There is always a magnetic field around a wire carrying a current in only one direction. The wires in our houses are not magnetic because the current constantly changes direction.

11
Center
of Gravity

70. Balancing Blob

- *Cut a 12-inch piece of string. Tie a washer to one end and a large-headed tack to the other.*
- *Draw a blob design of your own creation on a piece of stiff cardboard. Cut out your design.*
- *Glue a penny to the edge of the blob.*
- *Use a nail to punch 5 holes around the edge of the blob. Try to space the holes evenly.*
- *Stick the tack through one of the holes and into a tackboard to hold it. If you do not have a tackboard, use a cardboard box.*
- *Allow the blob to hang on the tack with the string hanging straight down.*
- *Draw a line across the blob next to the string.*
- *Move the tack to another hole, and again draw a line along the string.*
- *Continue until you have used all 5 holes.*

Results There will be a place where all the lines cross or at least are very near to each other. Where they cross is called the *center of gravity*.

Why? The center of gravity is the point where the weight of the object is evenly distributed all around it. The pull of gravity caused the weighted string to hang straight down. The center of gravity spot of the blob was also pulled straight down. This made the center of gravity of the blob touch some point on the string every time it was moved.

Try This Place the blob's point of balance, the center of gravity, on the tip of your finger, and see if it balances.

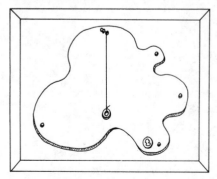
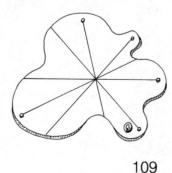

109

71. Over the Edge

- *Use 4 or 5 thin books.*
- *Lay the largest book on the table with about 1 inch hanging over the edge.*
- *Place a book on top of the first one with more of it hanging over the edge of the table.*
- *Continue adding the books in step-fashion.*
- *The top book is to be totally over the table's edge.*

Results The stack of books is balanced on the edge of the table, with the top book over the edge.

Why? The weight of the books is being pulled down by gravity. The portion of the weight directly over the table is more than that of the part hanging over the edge, so the books remain balanced.

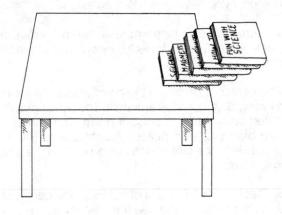

72. Big Foot

- *Stand so that your heels are against a wall.*
- *Slowly lean forward.*

Results You can lean forward only a short distance without losing your balance.

Why? As you lean forward, you remain balanced until your body's center of gravity passes the end of your toes, and then you start to fall. A person with long feet and a low center of gravity can lean farther out before losing his or her balance.

The leaning Tower of Pisa will fall if its center of gravity ever extends over its base.

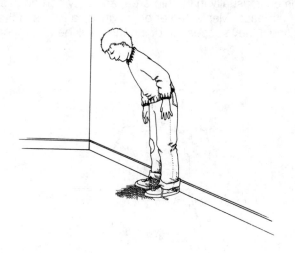

73. **Who's Stronger?**

- *Take 3 steps back from a wall.*
- *Have someone place a chair between you and the wall.*
- *Lean over, and place your head against the wall. Your legs should be at about a 45° angle to your torso.*
- *Holding to the edge of the seat of the chair, pick it up and hold the seat against your chest.*
- *Keeping the chair against your chest, stand up.*

Results Some people can do this with little or no effort, while others cannot do it at all.

Why? For those who have a low center of gravity, the weight of the chair will not keep them from standing up. The center of gravity for women is usually in the hip area, and they can do this experiment with ease. Men's center of gravity as a rule is in the upper torso, and if they succeed in doing this it will be with much effort.

74. Magic Box

- *Place a rock in the inside corner of a cigar box.*
- *Close the lid.*
- *Set the corner of the box with the rock on the edge of the table. Be sure that no part of the rock is over the table's edge.*

Results The box balances, with most of it suspended over the edge of the table.

Why? The weight of the rock makes the center of gravity of the box and its contents over the table top. This just means that more weight is pushing down on the table than is hanging over the edge.

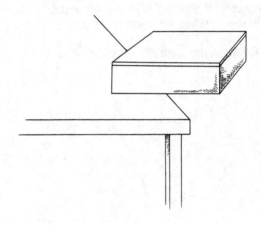

75. Butterfly Magic

- *Trace this butterfly on a piece of paper, and cut out the design.*
- *Use the paper butterfly as a pattern to draw the design on stiff cardboard.*
- *Cut the cardboard design out.*
- *Glue 2 pennies to the underside of the wings.*
- *Allow the glue to dry before you turn the butterfly over and decorate the top side.*
- *Place the magic butterfly's nose on the tip of your finger.*

Results The cardboard butterfly balances on its nose.

Why? The nose becomes the center of gravity with the pennies on the wing tips. If the butterfly will not exactly balance on its nose, move the pennies around until it does.

Try This Balance the butterfly on your nose.

76. **Balancing Point**

- Hold a spoon so that the bowl faces down.
- Slip the 2 center prongs of a fork into the spoon's bowl, leaving the first and fourth fork prongs extending over the bottom of the spoon.
- Slip the pointed end of a flat toothpick through the first fork prong, and balance the toothpick on the edge of a glass. You may have to adjust the position of the toothpick until it balances.
- Ignite the free end of the toothpick.

Results The toothpick burns up to the rim of the glass. The fork and spoon balance, with just the charred tip of the toothpick touching the rim of the glass.

Why? Because of the angle of the handles of the spoon and fork, the charred toothpick tip is the point of center of gravity. Changing the angle of the handles would redistribute the weight; the center of gravity would change, and they would no longer balance on the toothpick tip.

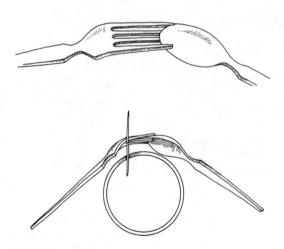

77. Mystifying Nails

■ *Seven nails are needed; #16 penny is a good size.*
■ *Cut a 2-inch by 4-inch wooden board about 8-inches long.*
■ *Hammer 1 of the nails into the center of the wooden block. You want the nail to be straight, tall, and secure.*
■ *The remaining 6 nails will be positioned horizontally on a table top: (a) Lay 1 nail on the table. (b) Place 4 additional nails perpendicular to the first nail. (c) The sixth nail goes across the top. Its head must point in the direction opposite that of the head of the bottom nail.*
■ *Carefully pick up the 6 nails, and balance the lower horizontal nail on the nailhead sticking out of the wooden block.*

Results The nails will balance.

Why? The hanging nails apply pressure to the 2 horizontal nails. This pressure pulls them together so that they act as a single unit. The balancing is due to the equal distribution of weight around the support nail.

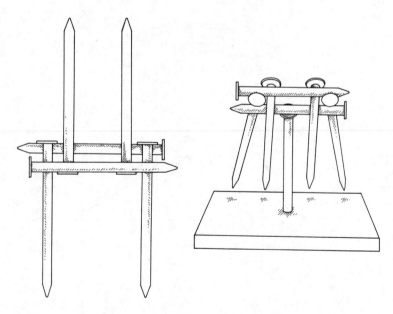

12
Inertia

78. Spinning Eggs

■ *Allow a hard-boiled and a raw egg to stand at room temperature for 20 minutes.*
■ *Place each egg on its side, and try to spin it.*

Results The raw egg does not spin as well as does the hard-boiled one.

Why? *Inertia* means an object in motion will continue to move until something stops it, and an object that is still remains so until something moves it.

The shell of the raw egg was spun by your hand; but the inertia of the liquid content kept it stationary. The liquid moves some, but very little compared to the shell. The lack of motion of the liquid forces the shell to stop spinning.

The boiled egg's shell and solid contents all spin together. The inertia of motion keeps it spinning until the friction of the table and air molecules stops it.

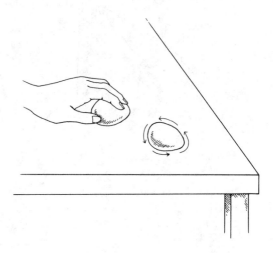

79. Snap!

- *Use a jar with an opening about 1½-inches in diameter.*
- *Place an index card over the mouth of the jar.*
- *Position a dime on top of the card so that it is centered above the jar's mouth.*
- *Use your fingers to snap the card, causing it to move quickly forward.*

Results The dime will fall into the jar as the card moves forward.

Why? The card moves because you hit it. The stationary inertia of the dime keeps it from moving with the card. Gravity pulls it straight down into the jar as soon as the card no longer supports it.

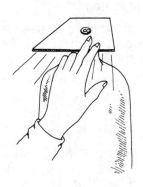

80. **Zip!**

- Cut a strip of wax paper 3-inches by 12-inches.
- Lay the strip on a table, and place a soda bottle over the end.
- Holding the free end, push the paper close to the bottle.
- Pull the paper away from the bottle with a straight, <u>quick</u>, forceful motion.

Results The paper moves from under the bottle, but the bottle remains in the same place.

Why? Because of inertia, the bottle does not move with the paper.

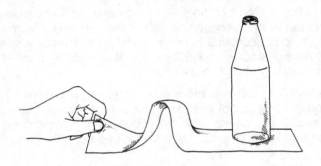

81.Penny Snap

- *Stack at least 4 pennies neatly on a table.*
- *Position another penny about 2-inches from the stack.*
- *Snap the single penny with your finger so that it hits the bottom penny in the stack.*

Results There are several possible results:

- The bottom penny from the stack moves forward, and the penny you snapped goes off at an angle to the stack.
- The bottom penny in the stack moves forward, and the penny you snapped stops, bounces back, or replaces the previous bottom penny on the bottom of the stack.

Why? There is an exchange of energy and inertia between the resting penny on the bottom and the moving penny. The moving penny tends to continue to move, and if it hits the stack straight on, it usually stops or bounces back. It sometimes pushes the bottom penny out and takes its place.

The snapped penny will move at an angle to the stack if it hits it off center. Only part of its energy is transferred in this type of hit. The speed and direction of the bottom penny depend on the speed and angle of impact of the snapped penny.

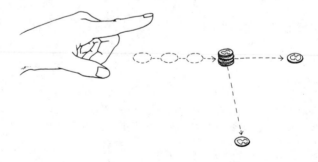

82. Thump!

- *Use a ruler that has a center groove. Lay it on a table.*
- *Place 5 marbles in the groove. Push them tightly together.*
- *Place 1 marble about 1 inch from the group, and thump it so that it rolls forward and hits the end marble.*

Results The thumped marble stops when it strikes the end marble, and a marble from the opposite end moves forward.

- *Try again, using 2 marbles. Roll them toward the group by thumping the one on the outside.*

Results The marbles move together and stop when the end marble is struck. Two marbles move forward from the opposite end.

Why? The stationary marbles have inertia at rest, a tendency to stay motionless. The rolling marble(s) has inertia of motion, a tendency to keep moving. When the moving marble(s) strikes the stationary ones, there is an exchange of inertia; the moving one(s) stops, and a marble at rest starts rolling.

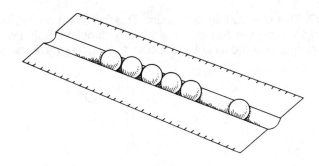

83. Sand Design

- *Secure a curtain push rod in a doorway, about 3 feet from the floor.*
- *Cut a string about 3 feet long. Tie both ends to the curtain rod as far apart as possible. Tape the ends to the rod to prevent their movement.*
- *Tie a loop of string around the top of a paper cup. Tie the ends of a string 12 inches long to opposite sides of this loop to form a handle.*
- *Attach the string handle on the cup to the hanging string from the curtain rod with a 12-inch string.*
- *Move the curtain rod so that the bottom of the cup is 3 to 6 inches from the floor.*
- *Place newspaper on the floor.*
- *Under the cup and on top of the newspaper, place a large piece of dark construction paper.*
- *Fill the cup with clean, dry sand.*
- *Punch a hole in the bottom of the cup with a pencil.*
- *Pull the cup back, and release it so that it swings freely.*

Results The sand pours out, forming a pattern of overlapping circles on the paper.

Why? This is a double pendulum. The top string moves in one direction, while the cup can move in any direction. Both swing in a pendulumlike motion of back and forth. They have inertia of

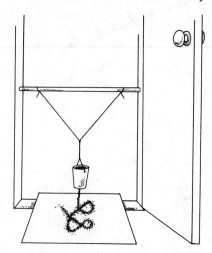

motion but in different directions. Since they are connected, there is an exchange of energy. The top pendulum cannot change the direction of its swing, so the exchange of energy causes the bottom, free-swinging cup to move in different directions.

84. Stop and Go

- *Stand 2 chairs back to back, 3 feet apart.*
- *Tie the ends of a string to the chairs to make a taut line.*
Make the line as tight as possible without pulling the chairs off balance.
- *Cut 2 additional pieces of string, each about 2 feet long, and tie a washer to the end of each string.*
- *Tie the free ends of the weighted strings about 1 foot apart to the line between the chairs.*
- *Start one of the weighted strings to swinging while the other hangs.*

Results In a short time the swinging weight will start slowing down, and the second weight will start to swing. The first weight will finally stop, and the weight originally at rest will be swinging. This process continues to reverse itself.

Why? The line both weighted strings are connected to transfers the energy back and forth. The swinging weight pulls on the line, which pulls on the resting weight until it starts to swing. They finally stop because of friction.

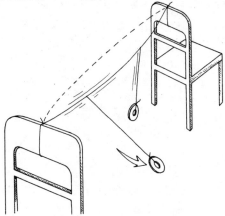

85. Uphill Roller

- *Stick a lump of clay on the inside rim of a metal coffee can.*
- *Make an incline by raising a large book about 1 inch on one end.*
- *Set the can on the bottom edge of the book with the clay at the top and just slightly toward the upper end of the incline.*
- *Place the lid on the can.*
- *Turn it loose.*

Results The can rolls up the incline.

Why? Gravity is pulling the clay straight down toward the upper incline, forcing the can to roll uphill.

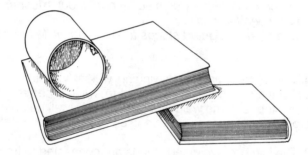

86. Wizard Stick

■ *Use a utility knife to cut notches out of a pencil. Start 2 inches from the eraser end, and cut as many notches as you can in the next 3 inches of the pencil.*
■ *Cut a ¾-inch by 2-inch propellor from thick cardboard. Stick a short straight pin through the center, and hollow out the hole so that the paper freely spins around the pin.*
■ *Stick the pin in the end of the eraser. You now have a wizard stick.*
■ *Hold the wizard stick in one hand and a round pen with the other hand.*
■ *Rub the round pen back and forth across the top of the notches. At the same time allow your thumb to rub against the side of the stick.*

Results The propellor turns away from the side you rubbed.

■ *Rub the notches again as before, but allow your index finger to rub against the opposite side this time.*

Results The propellor turns in the opposite direction. It again turns away from the side that was rubbed.

Why? Your finger directs the vibrating energy down the side of the stick, causing the propellor to spin. Changing the direction of the energy changes the direction of the spin.

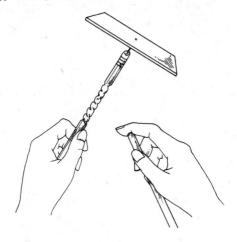

13
Leaping Lizards

15
Leaping Lizards

87.Leaping Lizards

■ Cut a strip of paper ½ inch by 6 inches.
■ Fold one end forward about ½ inch.
■ Fold this doubled end backward.
■ Continue to fold the paper forward and backward like an accordian.
■ Place a bowl filled with water on a table.
■ Sit so that the surface of the water is at eye level.
■ Wet <u>only one</u> of the end folds.
■ Hold the other end, and bring the wet fold close to the surface of the water.

Results The paper will leap to the water.

Why? Water molecules have a positive and a negative side. The positive side of one molecule attracts the negative side of another. This is a weak attraction, but strong enough to pull the paper into the water. Folding the strip makes it easier to move.

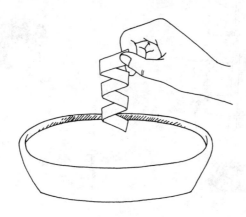

88. Hula Skirt

- *Fill a glass with water, and place it on a table.*
- *Plunge the bristles of an art brush into the water.*
- *Jiggle the bristles up and down in the water.*

Results The bristles move around like a hula skirt.

- *Pull the brush out of the water, and jiggle it up and down.*

Results The bristles stick together, and no amount of jiggling seems to separate them.

Why? Under the water, there is no surface film of water on the bristles. When they are pulled out of the glass, a thin layer of water sticks to their outer surface. The molecules in these layers are attracted to each other and tend to pull the bristles together.

89. Spiral Whirl

■ *From lightweight paper cut a spiral that has about a 2-inch diameter.*
■ *Fill a large bowl or pan with water.*
■ *Place a tiny dot of liquid dish soap on the indicated spot on the spiral.*
■ *Lay the spiral flat on the water's surface, soap side down.*

Results The paper will spin around.

Why? See the *Soap Boat* experiment for details on surface tension. The soap dissolves, breaking the surface tension and causing the water molecules to move away from the soapy area. The shape of the paper causes it to be pulled in different directions by the moving water; thus the spinning motion.

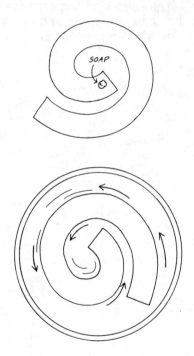

90. Soap Boat

■ *Cut a triangle, about 1½ inches high, from cardboard. Cut a small notch in the center of the short side of the paper triangle.*
■ *Fill a large bowl with water.*
■ *Place the boat in the water with the rear of the boat against the side.*
■ *Place 1 drop of liquid dish soap in the cut-out section of the boat.*

Results The boat races across the water.

Why? The attraction between the surface water molecules draws them close together. This produces a surface tension almost like a very thin skin across the water. The molecules are pulling on each other with equal force in all directions. The soap breaks the surface tension, which causes an unbalanced pull on the water molecules near the soap. The molecules are pulled away from the soap drop along with the paper boat. If the boat were very heavy, it would sink.

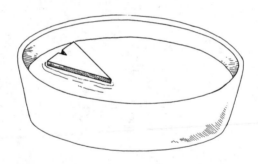

91. No Spills

■ *Fold a 3-inch by 6-inch piece of netting in half, or use a single layer of cheesecloth.*
■ *Fill a 10-ounce glass soda bottle to overflowing with water.*
■ *Stretch the cloth over the mouth of the bottle, and tightly secure it with a rubber band.*
■ *Pour more water into the bottle through the cloth. You want the bottle full and the cloth wet.*
■ *Quickly turn the bottle completely upside down. Just in case, you might want to do this over the sink.*

Results The water stays in the bottle. If you tilt it, it will pour out.

Why? The tiny holes in the cloth fill up with water. The water molecules have a strong enough attraction for each other to support the small amount of water above the mouth. If a jar with a larger mouth is used, the water will pour out.

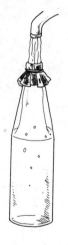

14

Did You Know?

92. How Light Travels

- *Cut a hole about the size of a dime in the center of 2 pieces of cardboard.*
- *Secure the cardboard pieces in an upright position with modeling clay. The pieces need to be about 4 inches apart, and the holes must be lined up so that you can look through both of them in this position.*
- *Light a candle. The candle must be tall enough for the flame to be the height of the holes in the cardboard.*
- *Hold a piece of white paper about 2 inches from the cardboard farthest from the candle.*

Results The light from the candle passes through the holes in both cardboard pieces and strikes the paper screen. If this does not happen, you need to line the holes up so that the light can pass through them.

- *Move one of the cardboard pieces about an inch to the left.*

Results The light cannot pass through the holes unless they are lined up.

Why? Light only travels in a straight line and cannot bend around corners. The holes have to be in a straight line for the light to pass through them.

Interesting Facts About Light
- Light can travel through empty space, but sound cannot.
- Light travels at a speed of 186,000 miles per second.
- Light can travel around the earth, which is 25,000 miles in diameter, in just 1/7 of a second.
- When you see a lightning flash, count the seconds until you hear the thunder. Multiply the number of seconds by 1,100 feet; that will be how far away the lightning was from you. One mile equals 5,280 feet. Figure how many miles the lightning flash was from you.

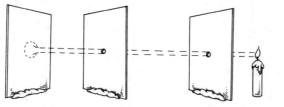

139

93. What Makes Shadows?

■ *Use a curtain push rod to hang a white sheet across an open doorway. This will be your screen.*
■ *Place a lamp about 6 feet behind the sheet.*
■ *Darken the room except for the lamp.*
■ *Hold your hands between the lamp and the screen to form images on the screen.*
■ *Your audience on the other side of the sheet will see shadows of the shapes made by your hands. (The position of the hands shown will produce an image of a dog.)*
■ *Experiment to form different animals, or put on a mime show.*
■ *To produce a larger shadow, stand closer to the light. Stand farther away to make a smaller shadow.*

Results Only the shadow of your body or hands will be seen by the audience on the other side of the curtain.

Why? A shadow is formed by blocking out rays of light. Two of the most famous shadows are formed by the earth and moon and called, respectively, *lunar* and *solar eclipses*.

A dark rain cloud is dark because the excess amount of water in it blocks out the sun's rays. The sun is always shining. It is the water in the clouds that makes the shadow.

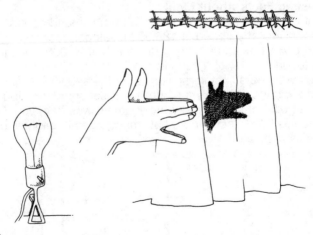

94. Light from Candy?

■ *This experiment must be done in complete darkness. You might use a closet with the door closed and a towel to block out any light that might shine under the door bottom.*

■ *Before closing the closet door get ready by: (a) Laying newspapers on the floor to catch the broken candy pieces. (b) Placing several pieces of wintergreen hard candies in your pocket. (c) Placing one candy piece in the jaws of a good pair of pliers and holding them so that the candy is directly in front of and near your eyes.*

■ *Have someone close the closet door and put the towel along the bottom.*

■ *Holding the pliers close to your eyes, squeeze them as hard as possible. You may have to use both hands to apply enough pressure if it does not work with only one.*

Results There will be a brief, faint glow of light very near the candy. It certainly will not illuminate the closet. In fact, it is difficult to see because of its short duration and low intensity. You may have to try several times for a good result. Pressure is the key to success in this experiment, and looking at the right spot is also important.

Why? Chemical energy is stored in everything, but only in some substances can this energy be changed to light energy by pressure. When the candy is squeezed, the molecules are pressed together, and the shifting around of the molecules allows the energy change.

It is believed that many of the UFO sightings in California are actually light flashes produced by the shifting of rocks along the fault line. When the rocks slip, there is a great amount of pressure produced. This pressure is much greater than that you produce with the pliers, so the light flashes are larger and more intense.

15
Toys That Teach Physics

95. Pinwheel

■ *Cut a 5-inch square from a sheet of notebook paper.*
■ *Draw 2 diagonal lines across the paper.*
■ *Use a nickel to draw a circle in the center.*
■ *Cut the 4 lines up to the circle. Four triangles connected in the center are formed.*
■ *Fold the left corner of one triangle to the center. Allow the point to overlap the center just a little.*
■ *Fold the left corner of each triangle to the center one at a time, allowing the points to slightly overlap the center.*
■ *Stick a straight pin through the center of the papers and into the side of a pencil eraser.*
■ *Move the pin around to hollow out the hole through the paper.*
■ *Blow on the pinwheel.*

Results It spins around.

Why? The wheel is designed to catch the air, which pushes the wheel around in a circular motion.

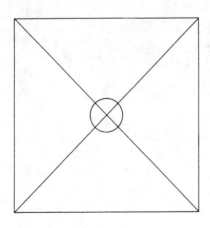

96. Papergyro

- *Cut 2 separate pieces of construction paper 3 inches wide and 4 inches long.*
- *Make a 3-inch cut down the center of each paper.*
- *Hold one of the pieces vertically in front of you.*
- *Bend the left wing forward and the right wing towards you.*
- *Hold the second paper, bending the left wing toward you and the right wing forward.*
- *Place a small paper clip in the center of the horizontal strip that makes the body of the papergyro.*
- *Hold each papergyro by its wings above your head and drop it.*

Results Both papergyros spin as they fall. One spins in a counterclockwise motion, and one spins clockwise.

Why? As the paper falls, air moves out from under each wing in all directions. The air molecules hit against the body of the papergyro as they try to escape in that direction. This happens under both wings, causing a rotating motion.

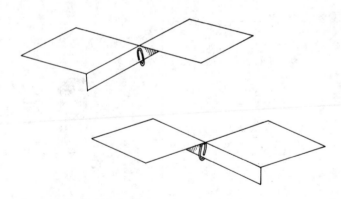

97. Boomerang

- *Use the pattern to cut a boomerang from stiff cardboard.*
- *Lay a piece of construction paper on top of a book.*
- *Place the boomerang on the paper as shown in the diagram.*
- *Tilt the book upward as far as possible without the boomerang falling.*
- *Direct your pencil along the side of the book toward the boomerang. You want to hit the boomerang so as to propel it upward in the direction the book is pointed. The greater the angle upward, the better the results.*

Results The boomerang will fly upward, turn, and fall back at your feet.

Why? The boomerang spins as it travels upward. As the upward speed decreases, the pull of gravity changes the direction of the spinning boomerang, bringing it back to your feet.

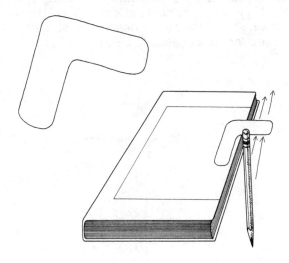

98. Parachute

- *Cut a 14-inch square from a sturdy plastic.*
- *Cut 4 pieces of string 14 inches long.*
- *Securely tape or tie a string to each corner of the plastic.*
- *Tie the free ends of the 4 strings together in a knot. Be sure the strings are all the same length.*
- *Tie a single string about 6 inches long to the knot.*
- *Add a weight, such as a washer, to the free end of the string.*
- *Pull the parachute up in the center. Squeeze the plastic to make it as flat as possible.*
- *Fold the parachute twice.*
- *Wrap the string loosely around the plastic.*
- *Throw the parachute up into the air.*

Results The parachute opens and slowly carries the weight to the ground.

Why? The weight falls first, unwinding the string because the parachute, being larger, is held back by the air. The air fills the plastic, slowing down the rate of descent. If the weight falls too quickly, a smaller object needs to be used.

99. Roller Can

■ *Punch 2 holes, ½ inch apart, in the center of the bottom of a metal coffee can. Do the same thing in the plastic lid of the coffee can.*

■ *Thread a rubber band through the 2 holes in the bottom of the can, leaving the looped ends inside the can. Do the same thing, using another rubber band, to the plastic lid, leaving the looped ends on the inside of the lid.*

■ *Keeping the rubber bands from twisting, bring all 4 end loops together inside the can, and secure them all together with a piece of string.*

■ *Tie a small fishing weight where the bands are connected. It is to hang down but must not touch the can.*

■ *Place the top on the can. The rubber bands must pull enough to hold the top on securely.*

■ *Roll the can on the floor and release.*

Results The can rolls forward, stops, and returns to you.

Why? The weight holds the rubber bands stationary in the center. The rolling can causes the bands to twist. The twisting bands apply force to the can, causing it to stop. The unwinding of the twisted bands starts the can rolling in the opposite direction.

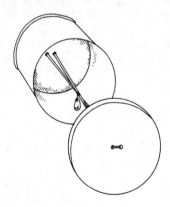

100. **Mousetrap Car**

- Cut a piece of wood or plywood 3 inches by 6 inches.
- Cut out a 1-inch square from one end.
- Tape a small mousetrap on top of the wood. The end of the mousetrap holding the long arming rod should be even with the edge of the 1-inch cut-out.
- Cut 2 4-inch long wooden dowels about the diameter of a pencil.
- Screw 4 eye hooks into the bottom of the wood. The eye of each hook must be large enough to allow a dowel to slip through it and turn with ease. Place the hooks ½ inch from the end and ½ inch from the side on all 4 corners.
- Use a nail to make holes in the center of 4 identical metal lids. Use lids approximately 2½ inches in diameter.
- From the outside, punch a small nail through the hole in one of the lids and into the end of a dowel rod.
- Thread the free end of the rod through 2 of the eye hooks on one end of the wood.
- Attach a lid to the free end of the rod with a small nail.
- Follow steps #7 through #9 to assemble the lids for the other end.
- Be sure that on the cut-out end the rod crosses under this opening. If not, reposition the eye hooks.
- Tie a string 12 inches long to the center of the back axle rod.
- Tie the free end of the string to the center of the mousetrap flapper.
- Turn the wheels backward to wind the string around the axle.
- After placing the car on the floor, set the mousetrap, and roll the car backward to take up any slack in the string.
- Trip the mousetrap.

Results The car quickly moves forward.

Why? As the flapper springs forward, the string is quickly pulled, rotating the axle and wheels.

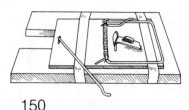
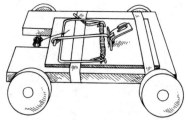

101. Hydraulic Lift

■ *Use 2 large syringes. Push the plunger in on one, and pull the plunger back on the other.*

■ *Connect the ends of the syringes with about 12 inches of plastic aquarium air tubing.*

■ *Cut the ends from a small box that is slightly taller than the syringes.*

■ *Cut a 1-inch section from one open end.*

■ *Use modeling clay to stand the closed syringe, plunger side up. Be sure not to crimp the air tube.*

■ *Slip the box over the standing syringe, fitting the cut out section over the tubing.*

■ *Place a rectangular piece of cardboard on top of the plunger of the standing syringe.*

■ *Set a small toy car in the center of the cardboard.*

■ *Push the plunger of the second syringe in slowly.*

Results The plunger of the standing syringe rises, lifting the car.

Why? Air is trapped inside both syringes and the tubing. When the open plunger is pushed in, the air is compressed and applies enough pressure on the closed plunger to push it and the toy car upward.

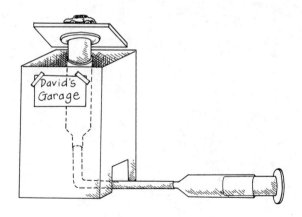

Appendix: Science Fair Projects

The purpose of a science fair project is to encourage the use of the scientific method in problem solving. Basically, the process requires you to identify a problem and state it in question form if possible. You must then plan experiments, make observations, and collect all information related to the problem. By studying this collection of data, an answer to the question can be formulated. This is called the conclusion to the experiment.

Broadening your knowledge of science is another goal of the project. A display that you whip together the night before presenting it defeats this purpose. But complex experiments with elaborate equipment are not a requirement for a blue ribbon entry, either. Evidence of solid research and of the ability to display your work rates high with the judges.

The display takes work, but it is a very important part of the presentation. You want the judges to be able to know exactly what your project is about and how much work you put into it just by *looking*. A 3-sided backboard is the best way to separate your work from that of others and gives you space to hang pictures, charts, graphs, etc. The size of this backboard is often regulated by local fair rules, so check before you start building. An average size is 4 feet high, with 3 separate, 2-feet wide panels. Wooden panels hinged together are sturdy and easy to work with, but a large cardboard box will do just as well. If the box is used, a support across the top will be needed.

The purpose, question, or problem to be solved can be placed on the left panel, the title at the top of the center panel, and the conclusion on the right panel. This leaves space for your printed materials to be displayed. Models and other large items can stand or lie on the table in front of the backboard.

Make the display neat, clean, and organized. Cut the letters out of colored construction paper for the needed headings. Remember that your display must be able to tell the judges what you have done.

You can write or print your question, research materials, and conclusion on paper, then secure it to the backboard. Do this *very neatly.* Draw most of your charts instead of using ones from a book. You learn more about the topic, and this is the most important objective of the project. Coloring the charts makes them much more visible and attractive. Take photographs of yourself performing experiments or just pictures of the different results. Display these with captions explaining each photo.

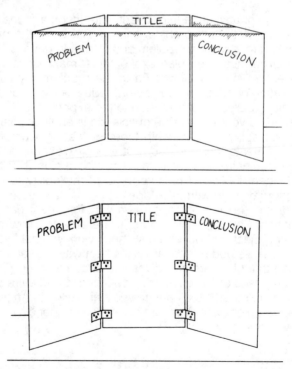

Selecting a Topic

Select a suitable topic to investigate. This means one that is acceptable to your teacher and interesting to you. Think ahead when choosing a subject, and ask yourself these questions:

- Can I find enough printed information about the topic to write my report?
- Will I be able to obtain the needed supplies to do the experiments?
- Do the science fair rules allow this type of experimentation?
- Can I display the model when it is finished?
- If electricity or other special conditions are needed, will they be available?

It is important to select a project that you can successfully complete and properly display. Every project requires some research. Projects including experiments and models make the experience more educational and also supply the much-needed display materials.

Do not feel that you must do something that has never been done before or find the cure for the common cold. New research is not the intention of an elementary science fair. Again, the objectives for the project are for you to learn more about science and problem-solving methods.

Any one of the experiments in this book can be used in designing your project. I will pick a few and show you how to take a topic and develop it into a suitable project. It is important to read all the topics in this section, since each one describes a different method or approach as well as additional suggestions.

The following are not intended to be complete instructions for a project but merely ideas to get you started. You are to take ideas such as these and plan your own project.

A. Title: *Why Airplanes Fly*
Question: What makes an airplane lift off the ground when other moving objects do not?

Notice that from the title a question has been formed. It is usually easier to work if you have a question to answer. Now you are ready to start your research. Check out books on airplanes, and keep the information that you read related to the question in a notebook. Be sure to write down the name of the book, the author, the publisher's name, and the date of publication as well as the pages you read. This information is used in writing a bibliography for your written report. It also helps you to find a book again if you are too sketchy in taking notes.

In reading about airplanes you will discover Bernoulli's principle, which explains that faster-moving air does not apply as much pressure to objects positioned perpendicular to the direction of the movement of the air as does slower-moving air. Now, that should impress almost anyone. As you perform the experiments in this book related to Bernoulli's principle, an explanation will be given in language that you can more readily understand. Do not include terms in your report that you do not really understand. Reword the explanations you find so that you understand them. You will not be able to explain the project if you cannot put the explanation in terms that *you* understand.

Draw pictures showing varied-shaped airplane wings and how the air flows over and under them. Construct paper airplanes using different patterns, and make drawings of their flight patterns. Display the drawings on the backboard, and set the paper models on the table area.

Do all the experiments in this book related to Bernoulli's principle and any others that you discover while researching this topic. The more you learn about the subject, the more confident you will be when talking to the project judges.

You are now ready to write a conclusion and organize the material for display. The conclusion must answer the question asked: *What makes an airplane lift off the ground when other moving objects do not?* Your answer can be in paragraph or outline form but must be specific. Airplanes fly because the uplift on the wings is greater than the drag. This is due to the decrease in pressure of the faster-moving air flow above the wings, etc. . . . Cars do not have enough uplift to fly because of the, etc. . . . Tell exactly why.

Each project needs a written report. It must include all the information you have learned about the topic, along with samples of the drawings. You will want to make small sketches of each of the drawings used on your display for the report. Start by writing the question at the top of the page, and include the data and research material collected. Your report should be more informative than the display. End the report with a conclusion which should be exactly the same as the one on your display. Every report needs a title page, and for best presentation put it in a folder.

Include a bibliography to show where you found your information. Study the resource list at the end of this book for the names of other science texts. It will also show you the correct form for writing a bibliography. Notice the different methods of listing texts and encyclopedia references.

B. Title: *Convection Currents*
Question: What is a convection current?
What effect do these currents have in nature?

This project has two questions. The first one could be answered too quickly so another related one is included. It increases the research needed to make the project more presentable and informative.

Smoking Chimney and *Spinning Spiral* are experiments in this text that can be used to answer the first question. Remember that you are not always allowed to demonstrate the experiments at the fair for the judges, so you want drawings and possibly photographs of your performing the experiments. The drawings will represent the experiment and its results. Be sure to include arrows to indicate the direction of the moving air or water.

As you read about convection currents, you will find material to answer the second question. The following are examples that could be used:

■ When the surface water on a pond cools in the winter, it sinks because the molecules get closer together and the water is heavier. The warmer water below the surface is lighter and rises. This keeps the nutrients in the water circulating. Ponds can become stagnant in the summer because the surface water heats up, and there is no circulation. Be sure to include the fact that it is due to the sinking of the heavier, cooler water in the northern oceans that there is such an abundant supply of sea organisms in the cold water. The sinking of the cold water and the rising of the warmer water bring the nutrients to the surface, providing food year round for the sea organisms.

The experiment *One Will, One Won't* can be used to demonstrate this. It would not be wise to display the bottles in this experiment, since they would most likely be knocked over, so be sure and take a photograph of the results or draw pictures showing what happened. If you are not familiar with photography, ask someone to help you. It is important that the lighting and background be right so that the glass bottles do not reflect too much light.

■ Sea and land breezes make a good example of convection currents, and pictures can be drawn to show the moving air currents. Water does not change its temperature as fast as does the land, so during the night the land cools faster, as does the air above it. On the beach at night the wind blows out toward the sea. This is because the warm air above the water rises, and the cooler land air rushes in to take its place, producing land breezes. A drawing with a person standing on the beach, his or her hair blowing toward the water, would be good.

■ During the day the land heats faster than does the water, and the air above the land rises, allowing the cooler sea air to rush in. This causes the cool sea breezes during the day. Another picture with a person standing on the beach can be used, but this time have his or her hair blowing toward the land.

These are just two examples, and you will find many more to use as you read about convection currents. If you are alert, you will observe effects of convection currents. Look for birds gliding on the rising warm air currents during the summer.

C. Title: *Lenses*
Question: How do different-shaped lenses change the path of light?

The experiments *Where's The Penny* and *Broken Pencil* make a good introduction for this project. They show that the direction of light is changed when the light passes from one substance into another. Make drawings to show the refraction of the light. A glass of water with a pencil in it could also be part of your display.

All the lens experiments are applicable to this project, but some would require special conditions for use as a display. The *Opaque Projector* is interesting but requires electricity. Check to see that this is available before starting. The *Camera Obscura* requires a darkened room. These and other experiments can be performed and drawings used instead of a working model. It is best to perform as many experiments as possible and include them in your written report, but choose for display the experiments that allow you to make models that do not require special conditions for operation.

An idea for this project that would be interesting would be to include drawings showing how different-shaped lenses are used to correct different vision disorders such as farsightedness and nearsightedness.

D. Title: *Weather Predictions*
Question: Can the changes of barometric pressure be used to predict weather conditions?

This project will require daily observations. Making a homemade barometer will be the first step. Follow the instructions given for the experiment *Barometer* to make your barometer. It will not give accurate readings, but it can indicate a rise and fall in pressure. Keep a daily record of the readings from your homemade barometer, the type of clouds present, temperature, wind, and humidity. You can also include the accurate barometer readings each day. This information is given daily on television and is printed in the newspaper. Keep these records for 30 days. Study your collection of data, and write a conclusion about the changes in the weather that accompany a rise or fall of the barometric pressure. If the pressure starts to fall, make a prediction of the weather to be expected.

Two drawings of the homemade barometer are needed to show the position of the pointer with a rising pressure and a falling pressure. Drawings of cloud types with specific barometric readings can be included. Charts showing barometer changes and weather predictions would make a good display item.

E. Title: *Gas Pressure*
Question: What are some of the effects of changes in gas pressure?

Thermometers work, geysers blow, popcorn pops, vacuum machines clean, and many other events are directly related to changes in gas pressure. The effects of changes in gas pressure are many. As you start researching this topic, one idea will lead to another, so select those that interest you most and develop them. Too much material will cause the project to appear cluttered and disorganized. The following are ideas to stimulate your interest. As has been suggested, do not use all of the ideas, but rather choose a few and thoroughly develop them.

159

A thermometer's measurement of temperature can be explained by using the *Bottle Thermometer* experiment. It makes a good display, and if actual demonstrations are allowed, you can show how the colored water drop moves up the tube. Explain this movement by telling how the gas molecules expand as the heat from your hands is transferred to the gas inside the bottle. Drawings illustrating the water-drop movement, along with arrows pointing out the direction of movement and the quantity of gas pressure can be made. Use more arrows to show an increase in pressure. The bottle thermometer works because the heated air inside the container expands and pushes up on the water bubble with more pressure than the air above pushes down. A comparison can be made to thermometers that have a liquid such as mercury or alcohol in them. It is the liquid in these instruments that expands and rises when heated, not a gas as in the bottle thermometer. Be sure to point out that the water bubble in your bottle thermometer is not expanding but is being pushed up by the expanding air below it.

The idea of the gas expanding and pushing the water bubble up also explains how a geyser works. As the gases inside the earth heat up, they expand and force the water out of any available crack in the earth's surface. Drawings showing this movement of water upward through the earth's cracks can be used as part of your project display. Include arrows indicating that the pressure of the gas below the water is greater than is the air pressure above. The title of a drawing could be *Movement from High Pressure to Low Pressure Area.* Make several drawings to show that at first there is equal pressure above and below the water level. When the subsurface gas pressure exceeds the pressure above and is forceful enough to lift the weight of the water, then out spurts the water. Remember, the water sprays out because the pressure of the air is not great enough to keep it down.

The *Fountain Machine* and *Rising Water* experiments demonstrate the movement of water from an area of higher pressure to one of lower pressure. In both of these experiments, instead of an increase of the pressure on one side described in the preceding geyser explanation, the pressure has been *decreased* on one side. In the *Fountain Machine* the gas in the bottle expands and moves through the tube into the lower bottle. When the gas that is left cools, it contracts, creating a partial vacuum. There is a great difference between the pressure inside the bottle and the air pressure outside it. The air is pushing down on the surface of the water, and because of the low pressure inside the top bottle the water is pushed up the tube to fill the empty spaces. Drawings made to illustrate this movement should include some indication that gravity is pulling down, trying to keep the water in the lower container. The pressure of the air is great enough to overcome the pull of gravity, however, and the water flows uphill.

The *Rising Water* experiment works on the principle that materials move from an area of high pressure to an area of lower pressure. To extend this idea, a study about mercury-filled barometers and home vacuum cleaners can be included. Is the dirt being sucked into the cleaning bag? Think about the machine's name—*vacuum cleaner.* The dictionary defines the word *vacuum* as a space almost exhausted of air. This would mean that the cleaner bag, being a vacuum, is only partially filled with air. The pressure inside the bag is less than that outside the bag. The air in the room has a much greater pressure than that inside the bag and pushes the dirt into the cleaning bag.

Actually, nothing is ever sucked into a container; instead, it is pushed in. The experiment *Two Worse Than One?* demonstrates that to drink through a straw one lowers the air pressure within the mouth and the air in the room pushes the liquid up the straw and into the mouth.

A tornado also demonstrates the moving of materials from a high-pressure area to an area of low pressure. When a tornado nears a house, why does the building explode outward? Why are people cautioned to open windows when a tornado is expected in the area? This topic allows for interesting, informative, and helpful research. Drawings using arrows to show pressure inside and outside windows and walls can be used. Be sure to point out that on a normal day the pressure on the inside and outside of the structure is equal. The swirling winds of the tornado produce a very low pressure area outside the house. Open windows allow the high-pressure air inside to escape. The pressure inside the house does not increase but is much higher in comparison to the very low pressure of the air on the outside. The air pressure inside the house is strong enough to push the walls outward. The *Sonic Egg* experiment best demonstrates the effect caused by a low-pressure area. To further demonstrate the strength of air, use the *Heavy Air* experiment. The *Inverted Glass* experiment also demonstrates the force of air. Do not forget that the exploding of popcorn is also due to the changing of captured water drops into gas, and this expanding gas pushes with such force that the corn kernels explode outward.

Presentation

An oral presentation for your class is usually a requirement for a science fair project. It is at this time that the teacher evaluates the work done and chooses the entries for the local fair. *Do not* plan to *read* your report to the class. Regardless of the amount of work you have done in preparing this project, a poor presentation at this point can disqualify you. Prepare a speech. Make it short, and practice in front of a friend. If possible, tape your practice presentation and listen to it. Review your notes if there is to be a question-and-answer session after your presentation.

You can decide on how best to dress for a class presentation, but I suggest that for the local fair you make a special effort to look nice. You are representing your work. Actually, you are acting as a salesperson for your project, and you want to present the very best possible image. Your appearance shows how much personal pride you have in yourself, and that is the first step in introducing your product, your science project. Approach the presentation with enthusiasm and excitement, and your reward will be self-satisfaction in a job well done, and just maybe a first place ribbon. Good Luck!

Glossary

Air. A mixture of gases surrounding the earth. It contains about 78 percent nitrogen, 21 percent oxygen, and 1 percent other gases.

Archimedes (287–212 B.C.). A Greek scientist who discovered that all objects float due to an upward force exerted by the liquid in which the object is floating. This is called the *buoyant force* and is equal to the weight of the liquid displaced by the floating object.

Atom. The smallest part that an element can be divided into and retain its identity.

Bernoulli, Daniel (1700–1782). A Swiss mathematician who discovered that the faster a fluid (either gas or liquid) flows, the less pressure it exerts on objects around it. This is known as *Bernoulli's Principle*.

Boiling Point. The temperature that supplies the amount of heat needed to energize vapor bubbles inside a liquid enough so they can break through the liquid's surface.

Buoyancy. Being able to float on water or in air.

Buoyant Force. The upward force exerted by a liquid on any object in or on the liquid. If this force is great enough, the object is said to be buoyant or having buoyancy.

Center of Gravity. The point on an object where it will balance.

Charged. Having an excess of either positive or negative particles.

Compass. A device used to indicate direction with respect to the earth's magnetic north pole.

Compress. To apply pressure to a substance, and cause a decrease in size.

Concave. An inward curve.

Condense. To change from a gas to a liquid.

Contract. To get smaller in size.

Convection Currents. Movement of heated air or water upward and the replacement by cooler, heavier air or water.

Convex. Curved outward.

Deflate. To remove a gas; the size of the container gets smaller.

Displaced. To push out of the way and replace.

Electron. A negative particle spinning around the nucleus of an atom.

Evaporate. To change from a liquid to a gas.

Expand. To get larger.

Gravity. The pull the earth has on a substance near or on it.

Illuminate. To light up.

Inertia. The resistance to a change in motion or rest.

Inflate. To fill with a gas; the container increases in size.

Kindling Temperature. The temperature at which an object will burn.

Lens. A material through which light passes and the direction of the rays are changed.

Magnet. A material that has a magnetic field around it. It attracts iron, nickel, cobalt, or alloys of these metals.

Magnetic Field. The force field around a magnet. This field flows out of the north-seeking end and back into the south-seeking end.

Molecule. A unit consisting of 2 or more atoms. The smallest part a substance can be separated into without changing its identity. (For example, a glass of water contains millions of water molecules. One single molecule of water contains 2 atoms of hydrogen and 1 atom of oxygen bonded together.)

North Pole. In a magnet, the end seeking the magnetic north pole of the earth. The magnetic field flows out of this end of a magnet.

Oxygen. A gas necessary for plant and animal survival; makes up about 21 percent of the volume of air; necessary for burning.

Primary Colors. Red, blue, and yellow. All other colors are a combination of these.

Prism. A material that refracts light that passes through it.

Real Image. An image that can be projected onto a screen.

Recoil. An explosion in a closed container produces equal forces in all directions. Opening one end of the container produces an unbalanced force, and the container will move in the direction opposite the open end; this movement is called the *recoil.*

Reflect. To bounce back from a surface.

Refraction. The changing of the direction of light when it passes from one substance into another.

Regelation. The melting of ice due to an increase in pressure.

Rusting. The slow chemical combination of an element with oxygen. The most common rusting reaction is between iron and oxygen, producing a reddish powder, iron oxide.

South Pole. The end of a magnet seeking the south magnetic pole of the earth. The magnetic field flows from the north pole of a magnet and back into the south pole.

Static Electricity. A buildup of positive or negative charges on a substance; caused by the friction between materials.

Surface Tension. The force of attraction between the molecules on the surface of a liquid. It behaves as if there were a thin skin pulled across the liquid.

Symmetrical. A line drawn through the center of an object will produce two identical mirror images if the object is symmetrical.

Vapor. A gas.

Resource List

Amery, Heather, and Littler, Angela. *The Fun Craft Book of Magnets and Batteries.* New York: Scholastic Book Service, 1976.

Armstrong, H. A., and Newbury, N. F. *The Young Experimenter.* New York: Sterling Publishing Co., 1960.

Beck, Derek, and McNeil, Mary Jane. *The Fun Craft Book of Flying Models: Paper Planes That Really Fly.* New York: Scholastic Book Service, 1976.

Editors of the Young People's Science Encyclopedia. *Young People's Science Dictionary.* Chicago: Children's Press, Inc., 1964.

Jensen, Rosalie, and Spector, Deborah. *Teaching Mathematics to Young Children: A Basic Guide.* Englewood Cliffs, N.J.: Prentice-Hall, Inc., 1984.

Levenson, Elaine. *Teaching Children About Science: Ideas and Activities that Every Teacher and Parent Can Use.* Englewood Cliffs, N.J., 1985.

Lynde, Carleton John. *Science Experiments with Home Equipment.* Princeton, N.J.: C. Van Nostrand Co., Inc., 1949.

Raymo, Chet. *The Crust of Our Earth: An Armchair Traveler's Guide to the New Geology.* Englewood Cliffs, N.J.: Prentice-Hall, Inc., 1984.

————. *365 Starry Nights: An Introduction to Astronomy for Every Night of the Year.* Englewood Cliffs, N.J.: Prentice-Hall, Inc., 1983.

Sherrod, P. Clay, and Koed, Thomas L. *A Complete Manual of Amateur Astronomy.* Englewood Cliffs, N.J.: Prentice-Hall, Inc., 1981.

Sisson, Edith A. *Nature with Children of All Ages.* Englewood Cliffs, N.J.: Prentice-Hall, Inc., 1982.

Skolnick, J., et al. *How to Encourage Girls in Math and Science: A Guide for Parents and Teachers.* Englewood Cliffs, N.J.: Prentice-Hall, Inc., 1982.

Van Deman, Barry A., and McDonald, Ed. *Nuts and Bolts: A Matter-of-Fact Guide to Science Fair Projects.* Harwood Heights, Ill.: The Science Man Press, 1980.

Webster, David. *How to Do a Science Project.* New York: Franklin Watts, Inc., 1974.

Index